Ensemble Methods in Data Mining: Improving Accuracy Through Combining Predictions

Synthesis Lectures on Data Mining and Knowledge Discovery

Editor
RobertGrossman, *University of Illinois, Chicago*

Ensemble Methods in Data Mining: Improving Accuracy Through Combining Predictions
Giovanni Seni and John F. Elder
2010

Modeling and Data Mining in Blogosphere
Nitin Agarwal and Huan Liu
2009

Ensemble Methods in Data Mining: Improving Accuracy Through Combining Predictions Giovanni Seni and John F. Elder
ISBN: 978-3-031-00771-2   paperback   ISBN: 978-3-031-01899-2   ebook
DOI 10.1007/978-3-031-01899-2

A Publication in the Springer series
SYNTHESIS LECTURES ON DATA MINING AND KNOWLEDGE DISCOVERY

Lecture #2
Series Editor: Robert Grossman, *University of Illinois, Chicago* Series ISSN
Synthesis Lectures on Data Mining and Knowledge Discovery Print 2151-0067 Electronic 2151-0075

Ensemble Methods in Data Mining: Improving Accuracy Through Combining Predictions

Giovanni Seni
Elder Research, Inc. and Santa Clara University

John F. Elder
Elder Research, Inc. and University of Virginia

SYNTHESIS LECTURES ON DATA MINING AND KNOWLEDGE DISCOVERY #2

ABSTRACT

Ensemble methods have been called the most influential development in Data Mining and Machine Learning in the past decade. They combine multiple models into one usually more accurate than the best of its components. Ensembles can provide a critical boost to industrial challenges – from investment timing to drug discovery, and fraud detection to recommendation systems – where predictive accuracy is more vital than model interpretability.

Ensembles are useful with all modeling algorithms, but this book focuses on decision trees to explain them most clearly. After describing trees and their strengths and weaknesses, the authors provide an overview of regularization – today understood to be a key reason for the superior performance of modern ensembling algorithms. The book continues with a clear description of two recent developments: *Importance Sampling* (IS) and *Rule Ensembles* (RE). IS reveals classic ensemble methods – bagging, random forests, and boosting – to be special cases of a single algorithm, thereby showing how to improve their accuracy and speed. REs are linear rule models derived from decision tree ensembles. They are the most interpretable version of ensembles, which is essential to applications such as credit scoring and fault diagnosis. Lastly, the authors explain the paradox of how ensembles achieve greater accuracy on new data despite their (apparently much greater) complexity.

This book is aimed at novice and advanced analytic researchers and practitioners – especially in Engineering, Statistics, and Computer Science. Those with little exposure to ensembles will learn why and how to employ this breakthrough method, and advanced practitioners will gain insight into building even more powerful models. Throughout, snippets of code in R are provided to illustrate the algorithms described and to encourage the reader to try the techniques[1].

The authors are industry experts in data mining and machine learning who are also adjunct professors and popular speakers. Although early pioneers in discovering and using ensembles, they here distill and clarify the recent groundbreaking work of leading academics (such as Jerome Friedman) to bring the benefits of ensembles to practitioners.

The authors would appreciate hearing of errors in or suggested improvements to this book, and may be emailed at seni@datamininglab.com and elder@datamininglab.com. Errata and updates will be available from www.morganclaypool.com

KEYWORDS

ensemble methods, rule ensembles, importance sampling, boosting, random forest, bagging, regularization, decision trees, data mining, machine learning, pattern recognition, model interpretation, model complexity, generalized degrees of freedom

[1]R is an Open Source Language and environment for data analysis and statistical modeling available through the Comprehensive R Archive Network (CRAN). The R system's library packages offer extensive functionality, and be downloaded form http://cran.r-project.org/ for many computing platforms. The CRAN web site also has pointers to tutorial and comprehensive documentation. A variety of excellent introductory books are also available; we particularly like *Introductory Statistics with R* by Peter Dalgaard and *Modern Applied Statistics with S* by W.N. Venables and B.D. Ripley.

To the loving memory of our fathers, Tito and Fletcher

Contents

Acknowledgments

We would like to thank the many people who contributed to the conception and completion of this project. Giovanni had the privilege of meeting with Jerry Friedman regularly to discuss many of the statistical concepts behind ensembles. Prof. Friedman's influence is deep. Bart Goethels and the organizers of ACM-KDD07 first welcomed our tutorial proposal on the topic. Tin Kam Ho favorably reviewed the book idea, Keith Bettinger offered many helpful suggestions on the manuscript, and Matt Strampe assisted with R code. The staff at Morgan & Claypool – especially executive editor Diane Cerra – were diligent and patient in turning the manuscript into a book. Finally, we would like to thank our families for their love and support.

Giovanni Seni and John F. Elder
January 2010

Foreword by Jaffray Woodriff

John Elder is a well-known expert in the field of statistical prediction. He is also a good friend who has mentored me about many techniques for mining complex data for useful information. I have been quite fortunate to collaborate with John on a variety of projects, and there must be a good reason that ensembles played the primary role each time.

I need to explain how we met, as ensembles are responsible! I spent my four years at the University of Virginia investigating the markets. My plan was to become an investment manager after I graduated. All I needed was a profitable technical style that fit my skills and personality (that's all!). After I graduated in 1991, I followed where the data led me during one particular caffeine-fueled, double all-nighter. In a fit of "crazed trial and error" brainstorming I stumbled upon the winning concept of creating one "super-model" from a large and diverse group of base predictive models.

After ten years of combining models for investment management, I decided to investigate where my ideas fit in the general academic body of work. I had moved back to Charlottesville after a stint as a proprietary trader on Wall Street, and I sought out a local expert in the field.

I found John's firm, Elder Research, on the web and hoped that they'd have the time to talk to a data mining novice. I quickly realized that John was not only a leading expert on statistical learning, but a very accomplished speaker popularizing these methods. Fortunately for me, he was curious to talk about prediction and my ideas. Early on, he pointed out that my multiple model method for investing described by the statistical prediction term, "ensemble."

John and I have worked together on interesting projects over the past decade. I teamed with Elder Research to compete in the KDD Cup in 2001. We wrote an extensive proposal for a government grant to fund the creation of ensemble-based research and software. In 2007 we joined up to compete against thousands of other teams on the Netflix Prize - achieving a third-place ranking at one point (thanks partly to simple ensembles). We even pulled a brainstorming all-nighter coding up our user rating model, which brought back fond memories of that initial breakthrough so many years before.

The practical implementations of ensemble methods are enormous. Most current implementations of them are quite primitive and this book will definitely raise the state of the art. Giovanni Seni's thorough mastery of the cutting-edge research and John Elder's practical experience have combined to make an extremely readable and useful book.

Looking forward, I can imagine software that allows users to seamlessly build ensembles in the manner, say, that skilled architects use CAD software to create design images. I expect that Giovanni and John will be at the forefront of developments in this area, and, if I am lucky, I will be involved as well.

Jaffray Woodriff
CEO, Quantitative Investment Management
Charlottesville, Virginia
January 2010

[Editor's note: Mr. Woodriff's investment firm has experienced consistently positive results, and has grown to be the largest hedge fund manager in the South-East U.S.]

Foreword by Tin Kam Ho

Fruitful solutions to a challenging task have often been found to come from combining an ensemble of experts. Yet for algorithmic solutions to a complex classification task, the utilities of ensembles were first witnessed only in the late 1980's, when the computing power began to support the exploration and deployment of a rich set of classification methods simultaneously. The next two decades saw more and more such approaches come into the research arena, and the development of several consistently successful strategies for ensemble generation and combination. Today, while a complete explanation of all the elements remains elusive, the ensemble methodology has become an indispensable tool for statistical learning. Every researcher and practitioner involved in predictive classification problems can benefit from a good understanding of what is available in this methodology.

This book by Seni and Elder provides a timely, concise introduction to this topic. After an intuitive, highly accessible sketch of the key concerns in predictive learning, the book takes the readers through a shortcut into the heart of the popular tree-based ensemble creation strategies, and follows that with a compact yet clear presentation of the developments in the frontiers of statistics, where active attempts are being made to explain and exploit the mysteries of ensembles through conventional statistical theory and methods. Throughout the book, the methodology is illustrated with varied real-life examples, and augmented with implementations in R-code for the readers to obtain first-hand experience. For practitioners, this handy reference opens the door to a good understanding of this rich set of tools that holds high promises for the challenging tasks they face. For researchers and students, it provides a succinct outline of the critically relevant pieces of the vast literature, and serves as an excellent summary for this important topic.

The development of ensemble methods is by no means complete. Among the most interesting open challenges are a more thorough understanding of the mathematical structures, mapping of the detailed conditions of applicability, finding scalable and interpretable implementations, dealing with incomplete or imbalanced training samples, and evolving models to adapt to environmental changes. It will be exciting to see this monograph encourage talented individuals to tackle these problems in the coming decades.

Tin Kam Ho
Bell Labs, Alcatel-Lucent
January 2010

CHAPTER 1

Ensembles Discovered

...and in a multitude of counselors there is safety.
Proverbs 24:6b

A wide variety of competing methods are available for inducing models from data, and their relative strengths are of keen interest. The comparative accuracy of popular algorithms depends strongly on the details of the problems addressed, as shown in Figure 1.1 (from Elder and Lee (1997)), which plots the relative out-of-sample error of five algorithms for six public-domain problems. Overall, neural network models did the best on this set of problems, but note that every algorithm scored best or next-to-best on at least two of the six data sets.

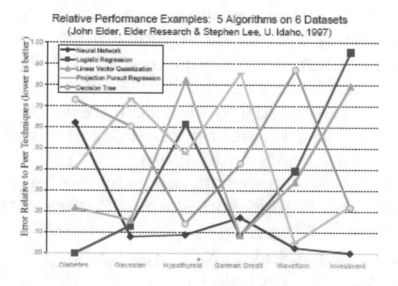

Figure 1.1: Relative out-of-sample error of five algorithms on six public-domain problems (based on Elder and Lee

(1997)).

How can we tell, ahead of time, which algorithm will excel for a given problem? Michie et al. (1994) addressed this question by executing a similar but larger study (23 algorithms on 22 data sets) and building a decision tree to predict the best algorithm to use given the properties of a data set[1]. Though the study was skewed toward trees—they were 9 of the 23 algorithms, and several of the (academic) data sets had unrealistic thresholds amenable to trees — the study did reveal useful lessons for algorithm selection (as highlighted in Elder, J. (1996a)).

Still, there is a way to improve model accuracy that is easier and more powerful than judicious algorithm selection: one can gather models into ensembles. Figure 1.2 reveals the out-of-sample accuracy of the models of Figure 1.1 when they are combined four different ways, including averaging, voting, and "advisor perceptrons" (Elder and Lee, 1997). While the ensemble technique of advisor perceptrons beats simple averaging on every problem, the difference is small compared to the difference between ensembles and the single models. Every ensemble method competes well here against the best of the individual algorithms.

This phenomenon was discovered by a handful of researchers, separately and simultaneously, to improve classification whether using decision trees (Ho, Hull, and Srihari, 1990), neural networks (Hansen and Salamon, 1990), or math theory (Kleinberg, E., 1990). The most influential early developments were by Breiman, L. (1996) with Bagging, and Freund and Shapire (1996) with AdaBoost (both described in Chapter 4).

One of us stumbled across the marvel of ensembling (which we called "model fusion" or "bundling") while striving to predict the species of bats from features of their echo-location signals (Elder, J., 1996b)[2]. We built the best model we could with each of several very different algorithms, such as decision trees, neural networks, polynomial networks, and nearest neighbors (see Nisbet et al. (2009) for algorithm descriptions). These methods employ different basis functions and training procedures, which causes their diverse surface forms – as shown in Figure 1.3 – and often leads to surprisingly different prediction vectors, even when the aggregate performance is very similar.

The project goal was to classify a bat's species noninvasively, by using only its "chirps." University of Illinois Urbana-Champaign biologists captured 19 bats, labeled each as one of 6 species, then recorded 98 signals, from which UIUC engineers calculated 35 time-frequency features[3]. Figure 1.4 illustrates a two-dimensional projection of the data where each class is represented by a different color and symbol. The data displays useful clustering but also much class overlap to contend with.

Each bat contributed 3 to 8 signals, and we realized that the set of signals from a given bat had to be kept together (in either training or evaluation data) to fairly test the model's ability to predict a species of an unknown bat. That is, any bat with a signal in the evaluation data must have no other signals from it in training. So, evaluating the performance of a model type consisted of building and cross-validating 19 models and accumulating the out-of-sample results (– a leave-one-*bat*-out method).

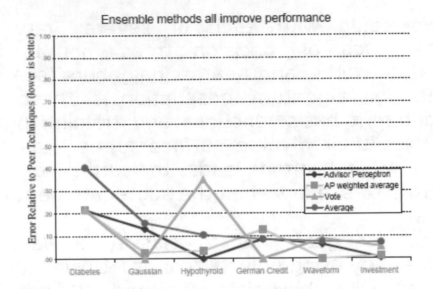

Figure 1.2: Relative out-of-sample error of four ensemble methods on the problems of Figure 1.1(based on Elder and Lee (1997)).

On evaluation, the baseline accuracy (always choosing the plurality class) was 27%. Decision trees got 46%, and a tree algorithm that was improved to look two-steps ahead to choose splits (Elder, J., 1996b) got 58%. Polynomial networks got 64%. The first neural networks tried achieved only 52%. However, unlike the other methods, neural networks don't select variables; when the inputs were then pruned in half to reduce redundancy and collinearity, neural networks improved to 63% accuracy. When the inputs were pruned further to be only the 8 variables the trees employed, neural networks improved to 69% accuracy out-of-sample. (This result is a clear demonstration of the need for regularization, as described in Chapter 3, to avoid overfit.) Lastly, nearest neighbors, using those same 8 variables for dimensions, matched the neural network score of 69%.

Despite their overall scores being identical, the two best models – neural network and nearest neighbor – disagreed a third of the time; that is, they made errors on very different regions of the data. We observed that the more confident of the two methods was right more often than not. (Their estimates were between 0 and 1 for a given class; the estimate more close to an extreme was usually more correct.) Thus, we tried averaging together the estimates of four of the methods – two-step decision tree, polynomial network, neural network, and nearest neighbor – and achieved 74% accuracy – the best of all. Further study of the lessons of each algorithm (such as when to ignore an estimate due to its inputs clearly being outside the algorithm's training domain) led to improvement reaching 80%. In short, it was discovered to be possible to break through the asymptotic performance ceiling of an individual algorithm by employing the estimates of multiple algorithms. Our fascination with what came to be known as ensembling began.

1.1 BUILDING ENSEMBLES

Building an ensemble consists of two steps: (1) constructing varied models and (2) combining their estimates (see Section 4.2). One may generate component models by, for instance, varying case weights, data values, guidance parameters, variable subsets, or partitions of the input space. Combination can be accomplished by voting, but is primarily done through model estimate weights, with gating and advisor perceptrons as special cases. For example, Bayesian model averaging sums estimates of possible models, weighted by their posterior evidence. Bagging (*bootsrap aggregating*; Breiman, L. (1996)) bootstraps the training data set (usually to build varied decision trees) and takes the majority vote or the average of their estimates (see Section 4.3). Random Forest (Ho, T., 1995; Breiman, L., 2001) adds a stochastic component to create more "diversity" among the trees being combined (see Section 4.4) AdaBoost (Freund and Shapire, 1996) and ARCing

(Breiman, L., 1996) iteratively build models by varying case weights (up-weighting cases with large current errors and down-weighting those accurately estimated) and employs the weighted sum of the estimates of the sequence of models (see Section 4.5). Gradient Boosting (Friedman, J., 1999, 2001) extended the AdaBoost algorithm to a variety of error functions for regression and classification (see Section 4.6).

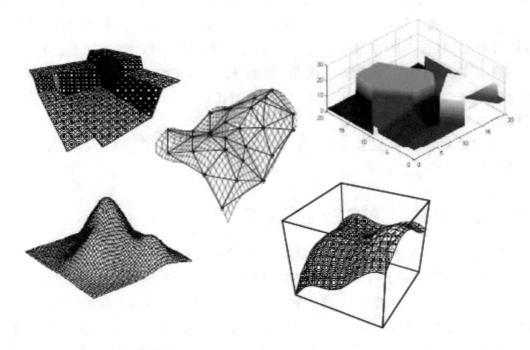

Figure 1.3: Example estimation surfaces for five modeling algorithms. Clockwise from top left: decision tree, Delaunay planes (based on Elder, J. (1993)), nearest neighbor, polynomial network (or neural network), kernel.

Figure 1.4: Sample projection of signals for 6 different bat species.

The Group Method of Data Handling (GMDH) (Ivakhenko, A., 1968) and its descendent, Polynomial Networks (Barron et al., 1984; Elder and Brown, 2000), can be thought of as early ensemble techniques. They build multiple layers of moderate-order polynomials, fit by linear regression, where variety arises from different variable sets being employed by each node. Their combination is nonlinear since the outputs of interior nodes are inputs to polynomial nodes in subsequent layers. Network construction is stopped by a simple cross-validation test (GMDH) or a complexity penalty. An early popular method, Stacking (Wolpert, D., 1992) employs neural networks as components (whose variety can stem from simply using different guidance parameters, such as initialization weights), combined in a linear regression trained on leave-1-out estimates from the networks.

Models have to be individually good to contribute to ensembling, and that requires knowing when to stop; that is, how to avoid overfit – the chief danger in model induction, as discussed next.

1.2 REGULARIZATION

A widely held principle in Statistical and Machine Learning model inference is that accuracy and simplicity are both desirable. But there is a tradeoff between the two: a flexible (more complex) model is often needed to achieve higher accuracy, but it is more susceptible to overfitting and less likely to generalize well. Regularization techniques "damp down" the flexibility of a model fitting procedure by augmenting the error function with a term that penalizes model complexity. Minimizing the augmented error criterion requires a certain increase in accuracy to "pay" for the increase in model complexity (e.g., adding another term to the model). Regularization is today understood to be one of the key reasons for the superior performance of modern ensembling algorithms.

An influential paper was Tibshirani's introduction of the Lasso regularization technique for linear models (Tibshirani, R., 1996). The Lasso uses the sum of the absolute value of the coefficients in the model as the penalty function and had roots in work done by Breiman on a coefficient post-processing technique which he had termed Garotte (Breiman et al., 1993).

Another important development came with the LARS algorithm by Efron et al. (2004), which allows for an efficient iterative calculation of the Lasso solution. More recently, Friedman published a technique called Path Seeker (PS) that allows combining the Lasso penalty with a variety of loss (error) functions (Friedman and Popescu, 2004), extending the original Lasso paper which was limited to the Least-Squares loss.

Careful comparison of the Lasso penalty with alternative penalty functions (e.g., using the sum of the squares of the coefficients) led to an understanding that the penalty function has two roles: controlling the "sparseness" of the solution (the number of coefficients that are non-zero) and controlling the magnitude of the non-zero coefficients ("shrinkage"). This led to development of the Elastic Net (Zou and Hastie, 2005) family of penalty functions which allow searching for the best shrinkage/sparseness tradeoff according to characteristics of the problem at hand (e.g., data size, number of input variables, correlation among these variables, etc.). The Coordinate Descent algorithm of Friedman et al. (2008) provides fast solutions for the Elastic Net.

Finally, an extension of the Elastic Net family to non-convex members producing sparser solutions (desirable when the number of variables is much larger than the number of observations) is now possible with the Generalized Path Seeker algorithm (Friedman, J., 2008).

1.3 REAL-WORLD EXAMPLES: CREDIT SCORING + THE NETFLIX CHALLENGE

Many of the examples we show are academic; they are either curiosities (bats) or kept very simple to best illustrate principles. We close Chapter 1 by illustrating that even simple ensembles can work in very challenging industrial applications. Figure 1.5 reveals the out-of-sample results of ensembling up to five different types of models on a credit scoring application. (The output of each model is ranked, those ranks are averaged and re-ranked, and the credit defaulters in a top percentage is counted. Thus, lower is better.) The combinations are ordered on the horizontal axis by the number of models used, and Figure 1.6 highlights the finding that the mean error reduces with increasing degree of combination. Note that the final model with all five component models does better than the best of the single models.

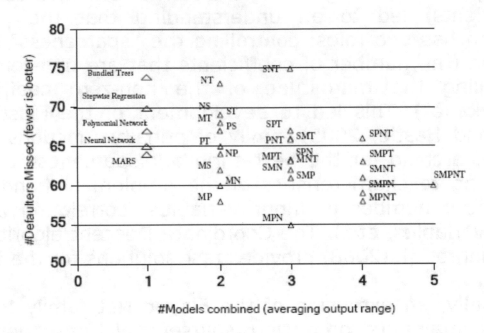

Figure 1.5: Out-of-sample errors on a credit scoring application when combining one to five different types of models into ensembles. *T* represents bagged trees; *S*, stepwise regression; *P*, polynomial networks; *N*, neural networks; *M*, MARS. The best model, *MPN*, thus averages the models built by MARS, a polynomial network, and a neural network algorithm.

Each model in the collection represents a great deal of work, and it was constructed by advocates of that modeling algorithm competing to beat the other methods. Here, MARS was the best and bagged trees was the worst of the five methods (though a considerable improvement over single trees, as also shown in many examples in Chapter 4).

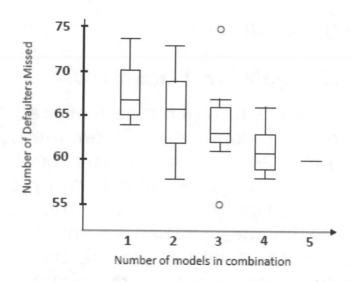

Figure 1.6: Box plot for Figure 1.5; median (and mean) error decreased as more models are combined.

Most of the ensembling being done in research and applications use variations of one kind of modeling method – particularly decision trees (as described in Chapter 2 and throughout this book). But one great example of heterogenous ensembling captured the imagination of the "geek" community recently. In the Netflix Prize, a contest ran for two years in which the first team to submit a model improving on Netflix's internal recommendation system by 10% would win $1,000,000. Contestants were supplied with entries from a huge movie/user matrix (only 2% non-missing) and asked to predict the ranking (from 1 to 5) of a set of the blank cells. A team one of us was on, *Ensemble Experts*, peaked at 3rd place at a time when over 20,000 teams had submitted. Moving that high in the rankings using ensembles may have inspired other leading competitors, since near the end of the contest, when the two top teams were extremely close to each other and to winning the prize, the final edge was obtained by weighing contributions from the models of up to 30 competitors.

Note that the ensembling techniques explained in this book are even more advanced than those employed in the final stages of the Netflix prize.

1.4 ORGANIZATION OF THIS BOOK

Chapter 2 presents the formal problem of predictive learning and details the most popular nonlinear method – decision trees, which are used throughout the book to illustrate concepts. Chapter 3 discusses model complexity and how regularizing complexity helps model selection. Regularization techniques play an essential role in modern ensembling. Chapters 4 and 5 are the heart of the book; there, the useful new concepts of Importance Sampling Learning Ensembles (ISLE) and Rule Ensembles – developed by J. Friedman and colleagues – are explained clearly. The ISLE framework allows us to view the classic ensemble methods of Bagging, Random Forest, AdaBoost, and Gradient Boosting as special cases of a single algorithm. This unified view clarifies the properties of these methods and suggests ways to improve their accuracy and speed. Rule Ensembles is a new ISLE-based model built by combining simple, readable rules. While maintaining (and often improving) the accuracy of the classic tree ensemble, the rule-based model is much more interpretable. Chapter 5 also illustrates recently proposed interpretation statistics, which are applicable to Rule Ensembles as well as to most other ensemble types. Chapter 6 concludes by explaining why ensembles generalize much better than their apparent complexity would seem to allow. Throughout, snippets of code in R are provided to illustrate the algorithms described.

[1]The researchers (Michie et al., 1994, Section 10.6) examined the results of one algorithm at a time and built a C4.5 decision tree (Quinlan, J., 1992) to separate those datasets where the algorithm was "applicable" (where it was within a tolerance of the best algorithm) to those where it was not. They also extracted rules from the tree models and used an expert system to adjudicate between conflicting rules to maximize net "information score." The book is online at http://www.amsta.leeds.ac.uk/charles/statlog/whole.pdf

[2]Thanks to collaboration with Doug Jones and his EE students at the University of Illinois, Urbana-Champaign.

[3]Features such as low frequency at the 3-decibel level, time position of the signal peak, and amplitude ratio of 1st and 2nd harmonics.

## CHAPTER 2

Predictive Learning and Decision Trees

In this chapter, we provide an overview of predictive learning and decision trees. Before introducing formal notation, consider a very simple data set represented by the following data matrix:

Table 2.1: Asimple data set. Each row represents a data "point" and each column corresponds to an "attribute." Sometimes, attribute values could be unknown or missing (denoted by a '?' below).

TI	PE	Response
1.0	M2	good
2.0	M1	bad
...	...	...
4.5	M5	?

Each row in the matrix represents an "observation" or data point. Each column corresponds to an attribute of the observations: *TI, PE*, and *Response*, in this example. *TI* is a numeric attribute, *PE* is an ordinal attribute, and *Response* is a categorical attribute. A categorical attribute is one that has two or more values, but there is no intrinsic ordering to the values – e.g., either good or bad in Table 2.1. An ordinal attribute is similar to a categorical one but with a clear ordering of the attribute values. Thus, in this example *M1* comes before *M2, M2* comes before *M3,* etc. Graphically, this data set can be represented by a simple two-dimensional plot with numeric attribute *TI* rendered on the horizontal axis and ordinal attribute *PE*, rendered on the vertical axis (Figure 2.1).

When presented with a data set such as the one above, there are two possible modeling tasks:

1.   Describe: Summarize   existing   data   in   an understandable and actionable way

2. Predict: What is the "Response" (e.g., class) of new point o ? See (Hastie et al., 2009).

More formally, we say we are given "training" data $D = \{y_i, x_{i1}, x_{i2}, \cdots, x_{in}\}_1^N = \{y_i, \mathbf{x}_i\}_1^N$ where

- y_i, x_{ij} are measured values of attributes (properties, characteristics) of an object

- y_i is the "response" (or output) variable

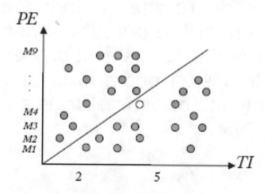

Figure 2.1: A graphical rendering of the data set from Table 2.1. Numeric and ordinal attributes make appropriate axes because they are ordered, while categorical attributes require color coding the points. The diagonal line represents

the best linear boundary separating the blue cases from the green cases.

- x_{ij} are the "predictor" (or input) variables
- $\mathbf{x}_i$ is the input "vector" made of all the attribute values for the i-th observation
- n is the number of attributes; thus, we also say that the "size" of $\mathbf{x}$ is n
- N is the number of observations
- D is a random sample from some unknown (joint) distribution $p(\mathbf{x}, y)$ – i.e., it is assumed there is a true underlying distribution out there, and that through a data collection effort, we've drawn a random sample from it.

Predictive Learning is the problem of using D to build a functional "model"

$$y = \hat{F}(x_1, x_2, \cdots, x_n) = \hat{F}(\mathbf{x})$$

which is the best predictor of y given input $\mathbf{x}$. It is also often desirable for the model to offer an interpretable description of how the inputs affect the outputs. When y is categorical, the problem is termed a "classification" problem; when y is numeric, the problem is termed a "regression" problem.

The simplest model, or estimator, is a linear model, with functional form

$$\hat{F}(\mathbf{x}) = a_0 + \sum_{j=1}^{n} a_j x_j$$

i.e., a weighted linear combination of the predictors. The coefficients $\{a_j\}_0^n$ are to be determined via a model fitting process such as "ordinary linear regression" (after assigning numeric labels to the points – i.e., +1 to the blue cases and

–1 to the green cases). We use the notation $\hat{F}_{(x)}$ to refer to the output of the fitting process – an approximation to the true but unknown function $F^*(\mathbf{x})$ linking the inputs to the output. The decision boundary for this model, the points where $\hat{F}_{(x)} = 0$, is a line (see Figure 2.1), or a plane, if $n > 2$. The classification rule simply checks which side of the boundary a given point is at – i.e.,

$$\hat{F}_{(x)} \begin{cases} \geq 0 & \text{(blue)} \\ \text{else} & \text{(green)} \end{cases}$$

In Figure 2.1, the linear model isn't very good, with several blue points on the (mostly) green side of the boundary.

Decision trees (Breiman et al., 1993; Quinlan, J., 1992) instead create a decision boundary by asking a sequence of nested yes/no questions. Figure 2.2 shows a decision tree for classifying the data of Table 2.1. The first, or root, node splits on variable TI: cases for which $T I \geq 5$, follow the left branch and are all classified as blue; cases for which $T I < 5$, go to the right "daughter" of the root node, where they are subject to additional split tests.

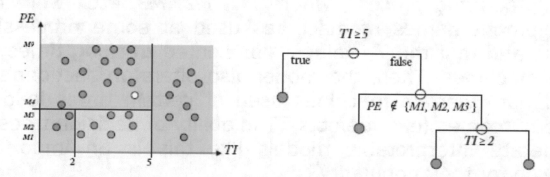

Figure 2.2: Decision tree example for the data of Table 2.1. There are two types of nodes: "split" and "terminal." Terminal nodes are given a class label. When reading the tree, we follow the left branch when a split test condition is met and the right branch otherwise.

At every new node the splitting algorithm takes a fresh look at the data that has arrived at it, and at all the variables and all the splits that are possible. When the data arriving at a given node is mostly of a single class, then the node is no longer split and is assigned a class label corresponding to the majority class within it; these nodes become "terminal" nodes.

To classify a new observation, such as the white dot in Figure 2.1, one simply navigates the tree starting at the top (root), following the left branch when a split test condition is met and the right branch otherwise, until arriving at a terminal node. The class label of the terminal node is returned as the tree prediction.

The tree of Figure 2.2 can also be expressed by the following "expert system" rule (assuming green = "bad" and blue = "good"):

$$TI \in [2, 5] \quad AND \quad PE \in \{M1, M2, M3\} \Rightarrow bad$$
$$ELSE\ good$$

which offers an understandable summary of the data (a descriptive model). Imagine this data came from a manufacturing process, where $M1$, $M2$, $M3$, etc., were the equipment names of machines used at some processing step, and that the TI values represented tracking times for the machines. Then, the model also offers an "actionable" summary: certain machines used at certain times lead to bad outcomes (e.g., defects). The ability of decision trees to generate interpretable models like this is an important reason for their popularity.

In summary, the predictive learning problem has the following components:

- *Data*: $D = \{y_i, x_i\}_1^N$

- *Model*: the underlying functional form sought from the data – e.g., a linear model, a decision tree model, etc.

We say the model represents a family $\mathcal{F}$ of functions, each indexed by a parameter vector **p**:

$$\hat{F}(\mathbf{x}) = \hat{F}(\mathbf{x}; \mathbf{p}) \in \mathcal{F}$$

In the case where $\mathcal{F}$ are decision trees, for example, the parameter vector **p** represents the splits defining each possible tree.

- *Score criterion:* judges the quality of a fitted model. This has two parts:

  ○ *Loss* function: Penalizes individual errors in prediction. Examples for regression tasks include the squared-error loss, $L(y, \hat{y}) = (y - \hat{y})^2$, and the absolute-error loss, $L(y, \hat{y}) = |y - \hat{y}|$. Examples for 2-class classification include the exponential loss, $L(y, \hat{y}) = \exp(-y \cdot \hat{y})$, and the (negative) binomial log-likelihood, $L(y, \hat{y}) = \log(1 + e^{-y \cdot \hat{y}})$.

  ○ *Risk:* the expected loss over all predictions, $R(\mathbf{p}) = E_{y,\mathbf{x}} L(y, F(\mathbf{x}; \mathbf{p}))$, which we often approximate by the average loss over the training data:

  $$\hat{R}(\mathbf{p}) = \frac{1}{N} \sum_{i=1}^{N} L(y_i, \hat{F}(\mathbf{x}_i; \mathbf{p})) \tag{2.1}$$

  In the case of ordinary linear regression (OLR), for instance, which uses squared-error

  $$\hat{R}(\mathbf{p}) = \hat{R}(\mathbf{a}) = \frac{1}{N} \sum_{i=1}^{N} \left(y_i - a_0 - \sum_{j=1}^{n} a_j x_j \right)^2$$

- *Search Strategy:* the procedure used to minimize the risk criterion – i.e., the means by which we solve

$$\hat{\mathbf{p}} = \arg\min_{\mathbf{p}} \hat{R}(\mathbf{p})$$

In the case of OLR, the search strategy corresponds to direct matrix algebra. In the case of trees, or neural networks, the search strategy is a heuristic iterative algorithm.

It should be pointed out that no model family is universally better; each has a class of target functions, sample size, signal-to-noise ratio, etc., for which it is best. For instance, trees work well when 100's of variables are available, but the output vector only depends on a few of them (say<10); the opposite is true for Neural Networks (Bishop, C., 1995) and Support Vector Machines (Scholkopf et al., 1999). How to choose the right model family then? We can do the following:

- Match the assumptions for particular model to what is known about the problem, or

- Try several models and choose the one that performs the best, or

- Use several models and allow each subresult to contribute to the final result (the ensemble method).

2.1 DECISIONTREE INDUCTION OVERVIEW

In this section, we look more closely at the algorithm for building decision trees. Figure 2.3 shows an example surface built by a regression tree. It's a piece-wise constant surface: there is a "region" R_m in input space for each terminal node in the tree – i.e., the (hyper) rectangles induced by tree cuts. There is a constant associated with each region, which represents the estimated prediction $\hat{y} = \hat{c}_m$ that the tree is making at each terminal node.

Formally, an *M*-terminal node tree model is expressed by:

$$\hat{y} = T(\mathbf{x}) = \sum_{m=1}^{M} \hat{c}_m I_{R_m}(\mathbf{x})$$

where $I_A(\mathbf{x})$ is 1 if $\mathbf{x} \in A$ and 0 otherwise. Because the regions are disjoint, every possible input $\mathbf{x}$ belongs in a single one, and the tree model can be thought of as the sum of all these regions.

Trees allow for different loss functions fairly easily. The two most used for regression problems are *squared-error* where the optimal constant $\hat{c}_m$ is the mean and the *absolute-error* where the optimal constant is the median of the data points within region R_m (Breiman et al., 1993).

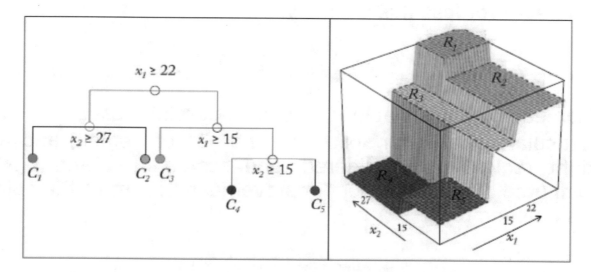

Figure 2.3: Sample regression tree and corresponding surface in input (**x**) space (adapted from (Hastie et al., 2001)).

If we choose to use squared-error loss, then the search problem, finding the tree $T(\mathbf{x})$ with lowest prediction risk, is stated:

$$\left\{ \hat{c}_m, \hat{R}_m \right\}_1^M = \operatorname*{arg\,min}_{\{c_m, R_m\}_1^M} \sum_{i=1}^N [y_i - T(x_i)]^2$$

$$= \operatorname*{arg\,min}_{\{c_m, R_m\}_1^M} \sum_{i=1}^N \left[y_i - \sum_{m=1}^M c_m I_{R_m}(x_i) \right]^2$$

To solve, one searches over the space of all possible constants and regions to minimize average loss. Unrestricted optimization with respect to $\{R_m\}_1^M$ is very difficult, so one universal technique is to restrict the shape of the regions (see Figure 2.4).

Joint optimization with respect to $\{R_m\}_1^M$ and $\{c_m\}_1^M$, simultaneously, is also extremely difficult, so a greedy iterative procedure is adopted (see Figure 2.5). The procedure starts with all the data points being in a single region R and computing a score for it; in the case of squared-error loss this is simply:

$$\hat{e}(R) = \frac{1}{N} \sum_{x \in R} \left(y_i - \operatorname{mean}\left(\{y_i\}_1^N\right) \right)^2$$

Then each input variable x_j, and each possible test s_j on that particular variable for splitting R into R^l (left region) and R^r (right region), is considered, and scores $\hat{e}(R^l)$ and $\hat{e}(R^r)$ computed. The quality, or "improvement," score of the split s_j is deemed to be

$$\hat{I}(x_j, s_j) = \hat{e}(R) - \hat{e}(R^l) - \hat{e}(R^r)$$

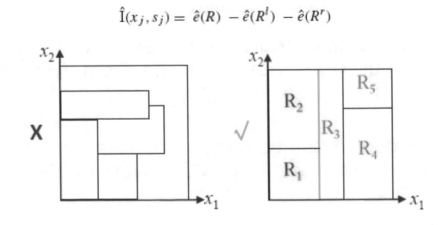

Figure 2.4: Examples of invalid and valid "regions" induced by decision trees. To make the problem of building a tree computationally fast, the region boundaries are restricted to be rectangles parallel to the axes. Resulting regions are simple, disjoint, and cover the input space (adapted from (Hastie et al., 2001)).

- Starting with a single region -- i.e., all given data
- At the m-th iteration:

```
for each region R with enough "impure" data
    for each attribute x_j in R
        for each possible split s_j of x_j
            record change in score when we partition R into R' and R'
    Choose (x_j , s_j ) giving maximum improvement
    Replace R with R'; add R'
```

Figure 2.5: Forward stagewise additive procedure for building decision trees.

i.e., the reduction in overall error as a result of the split. The algorithm chooses the variable and the split that improves the fit the most, with no regard to what's going to happen subsequently. And then the original region is replaced with the two new regions and the splitting process continues iteratively (recursively).

Note the data is 'consumed' exponentially—each split leads to solving two smaller subsequent problems. So, when should the algorithm stop? Clearly, if all the elements of the set $\{x : x \in R\}$ have the same value of y, then no split is going to improve the score – i.e., reduce the risk; in this case, we say the region R is "pure." One could also specify a maximum number of desired terminal nodes, maximum tree depth, or minimum node size. In the next chapter, we will discuss a more principled way of deciding the optimal tree size.

This simple algorithm can be coded in a few lines. But, of course, to handle real and categorical variables, missing values and various loss functions takes thousands of lines of code. In *R,* decision trees for regression and classification are available in the *rpart* package (rpart).

2.2 DECISION TREE PROPERTIES

As recently as 2007, a KDNuggets poll (Data Mining Methods, 2007) concluded that trees were the "method most frequently used" by practitioners. This is so because they have many desirable data mining properties. These are as follows:

1. *Ability to deal with irrelevant inputs.* Since at every node, we scan all the variables and pick the best, trees naturally do variable selection. And, thus, anything you can measure, you can allow as a candidate without worrying that they will unduly skew your results.

   Trees also provide a variable importance score based on the contribution to error (risk) reduction across all the splits in the tree (see Chapter 5).

2. *No data preprocessing needed.* Trees naturally handle numeric, binary, and categorical variables.

   Numeric attributes have splits of the form $x_j <$ *cut_value;* categorical attributes have splits of the form $x_j \in$ {*value*1, *value*2,...}.

   Monotonic transformations won't affect the splits, so you don't have problems with input outliers. If *cut_value* = 3 and a value x_j is 3.14 or 3,100, it's greater than 3, so it goes to the same side. Output outliers can still be influential, especially with squared-error as the loss.

3. *Scalable computation.* Trees are very fast to build and run compared to other iterative techniques. Building a

tree has approximate time complexity of $O(nN \log N)$.

4. *Missing value tolerant.* Trees do not suffer much loss of accuracy due to missing values.

   Some tree algorithms treat missing values as a separate categorical value. CART handles them via a clever mechanism termed "surrogate" splits (Breiman et al., 1993); these are substitute splits in case the first variable is unknown, which are selected based on their ability to approximate the splitting of the originally intended variable.

   One may alternatively create a new binary variable $x_{j\_}is\_NA$ (not available) when one believes that there may be information in x_j's being missing – i.e., that it may not be "missing at random."

5. *"Off-the-shelf" procedure: there are only few tunable parameters.* One can typically use them within minutes of learning about them.

6. *Interpretable model representation.* The binary tree graphic is very interpretable, at least to a few levels.

2.3 DECISION TREE LIMITATIONS

Despite their many desirable properties, trees also suffer from some severe limitations:

1. Discontinuous piecewise constant model. If one is trying to fit a trend, piecewise constants are a very poor way to do that (see Figure 2.6). In order to approximate a trend well, many splits would be needed, and in order to have many splits, a large data set is required.

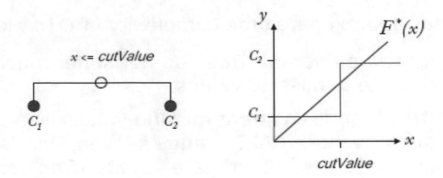

Figure 2.6: A 2-terminal node tree approximation to a linear function.

2. Data fragmentation. Each split reduces training data for subsequent splits. This is especially problematic in high dimensions where the data is already very sparse and can lead to overfit (as discussed in Chapter 6).

3. Not good for low "interaction" target functions $F^*(\mathbf{x})$. This is related to point 1 above. Consider that we can equivalently express a linear target as a sum of single-variable functions:

$$F^*(x) = a_o + \sum_{j=1}^{n} a_j x_j$$

$$= \sum_{j=1}^{n} f_j^*(x_j) \quad \text{i.e., no interactions, additive model}$$

and in order for x_j to enter the model, the tree must split on it, but once the root split variable is selected, additional variables enter as products of indicator functions. For instance, $\hat{R}_1$ in Figure 2.3 is defined by the product of $I(x_1 > 22)$ and $I(x_2 > 27)$.

4. Not good for target functions $F^*(\mathbf{x})$ that have dependence on many variables. This is related to point 2 above. Many variables imply that many splits are needed, but then we will run into the data fragmentation problem.

5. High variance caused by greedy search strategy (local optima) – i.e., small changes in the data (say due to sampling fluctuations) can cause big changes in the resulting tree. Furthermore, errors in upper splits are propagated down to affect all splits below it. As a result, very deep trees might be questionable.

Sometimes, the second tree following a data change may have a very similar performance to the first; this happens because typically in real data some variables are very correlated. So the end-estimated values might not be as different as the apparent difference by looking at the variables in the two trees.

Ensemble methods, discussed in Chapter 4, maintain tree advantages-except for perhaps interpretability-while dramatically increasing their accuracy. Techniques to improve the inter-pretability of ensemble methods are discussed in Chapter 5.

## CHAPTER 3

Model Complexity, Model Selection and Regularization

This chapter provides an overview of model complexity, model selection, and regularization. It is intended to help the reader develop an intuition for what *bias* and *variance* are; this is important because ensemble methods succeed by reducing bias, reducing variance, or finding a good tradeoff between the two. We will present a definition for regularization and see three different implementations of it. Regularization is a variance control technique which plays an essential role in modern ensembling. We will also review cross-validation which is used to estimate "meta" parameters introduced by the regularization process. We will see that finding the optimal value of these meta-parameters is equivalent to selecting the optimal model.

3.1 WHAT IS THE "RIGHT" SIZE OF A TREE?

We start by revisiting the question of how big to grow a tree, what is its right size? As illustrated in Figure 3.1, the dilemma is this: if the number of regions (terminal nodes) is too small, then the piecewise constant approximation is too crude. That intuitively leads to what is called "bias," and it creates error.

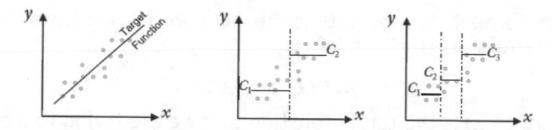

Figure 3.1: Representation of a tree model fit for simple 1-dimensional data. From left to right, a linear target function, a 2-terminal node tree approximation to this target function, and a 3-terminal node tree approximation. As the number of nodes in the tree grows, the approximation is less crude but overfitting can occur.

If, on the other hand, the tree is too large, with many terminal nodes, "overfitting" occurs. A tree can be grown all the way to having one terminal node for every single data point in the training data.[1] Such a tree will have zero error on the training data; however, if we were to obtain a second batch of data-test data-it is very unlikely that the original tree will perform as well on the new data. The tree will have fitted the noise as well as the signal in the training data-analogous to a child memorizing some particular examples without grasping the underlying concept.

With very flexible fitting procedures such as trees,we also have the situation where the variation among trees, fitted to different data samples from a single phenomenon, can be large. Consider a semiconductor manufacturing plant where for several consecutive days, it is possible to collect a data sample characterizing the devices being made. Imagine that a decision tree is fit to each sample to classify the defect-free vs. failed devices. It is the same process day to day, so one would expect the data distribution to be very similar. If, however, the trees are not very similar to each other, that is known as "variance."

3.2 BIAS-VARIANCE DECOMPOSITION

More formally, suppose that the data we have comes from the "additive error" model:

$$y = F^*(\mathbf{x}) + \varepsilon \quad (3.1)$$

where $F^*(\mathbf{x})$ is the target function that we are trying to learn. We don't really know F^*, and because either we are not measuring everything that is relevant, or we have problems with our measurement equipment, or what we measure has "noise" in it, the response variable we have contains the truth plus some error. We assume that these errors are independent and identically distributed. Specifically, we assume ε is normally distributed – i.e., $\varepsilon \sim N(0, \sigma^2)$ (although this is not strictly necessary).

Now consider the idealized aggregate estimator

$$\bar{F}(\mathbf{x}) = E\hat{F}_D(\mathbf{x}) \qquad (3.2)$$

which is the average fit over all possible data sets. One can think of the expectation operator as an averaging operator. Going back to the manufacturing example, each $\hat{F}$ represents the model fit to the data set from a given day. And assuming many such data sets can be collected, $\bar{F}$ can be created as the average of all those $\hat{F}$'s.

Now, let's look at what the error of one of these $\hat{F}$'s is on one particular data point, say $\mathbf{x}_0$, under one particular loss function, the squared-error loss, which allows easy analytical manipulation. The error, known as the Mean Square Error (MSE) in this case, at that particular point is the expectation of the squared difference between the target y and $\hat{F}$:

$$\begin{aligned}
\mathrm{Err}(\mathbf{x}_0) &= E\left[y - \hat{F}(\mathbf{x}) | \mathbf{x} = \mathbf{x}_0\right]^2 \\
&= E\left[F^*(\mathbf{x}_0) - \hat{F}(\mathbf{x}_0)\right]^2 + \sigma^2 \\
&= E\left[F^*(\mathbf{x}_0) - \bar{F}(\mathbf{x}_0) + \bar{F}(\mathbf{x}_0) - \hat{F}(\mathbf{x}_0)\right]^2 + \sigma^2
\end{aligned}$$

The derivation above follows from equations Equations (3.1) and (3.2), and properties of the expectation operator. Continuing, we arrive at:

$$= E\left[\bar{F}(x_0) - F^*(x_0)\right]^2 + E\left[\hat{F}(x_0) - \bar{F}(x_0)\right]^2 + \sigma^2$$
$$= \left[\bar{F}(x_0) - F^*(x_0)\right]^2 + E\left[\hat{F}(x_0) - \bar{F}(x_0)\right]^2 + \sigma^2$$
$$= \text{Bias}^2(\hat{F}(x_0)) + \text{Var}(\hat{F}(x_0)) + \sigma^2$$

The final expression says that the error is made of three components:

- $[\bar{F}(x_0) - F^*(x_0)]^2$: known as *squared-bias*, is the amount by which the average estimator $\bar{F}$ differs from the truth F^*. In practice, squared-bias can't be computed, but it's a useful theoretical concept.

- $E[\hat{F}(x_0) - \bar{F}(x_0)]^2$: known as *variance*, is the "spread" of the $\hat{F}$'s around their mean $\bar{F}$.

- σ^2: is the *irreducible error*, the error that was present in the original data, and cannot be reduced unless the data is expanded with new, more relevant, attributes, or the measurement equipment is improved, etc.

Figure 3.2 depicts the notions of squared-bias and variance graphically. The blue shaded area is indicative of the σ of the error. Each data set collected represents different realizations of the truth F^*, each resulting in a different y; the spread of these y's around F^* is represented by the blue circle. The model family $\mathcal{F}$, or model space, is represented by the region to the right of the red curve. For a given target realization y, one $\hat{F}$ is fit, which is the member from the model space $\mathcal{F}$ that is "closest" to y. After repeating the fitting process many times, the average $\bar{F}$ can be computed. Thus, the orange circle represents variance, the "spread" of the $\hat{F}$'s around their mean $\bar{F}$. Similarly, the "distance" between the average estimator $\bar{F}$ and the truth

F* represents model bias, the amount by which the average estimator differs from the truth.

Because bias and variance add up to MSE, they act as two opposing forces. If bias is reduced, variance will often increase, and vice versa. Figure 3.3 illustrates another aspect of this tradeoff between bias and variance. The horizontal axis corresponds to model "complexity." In the case of trees, for example, model complexity can be measured by the size of the tree.

At the origin, minimum complexity, there would be a tree of size one, namely a stump. At the other extreme of the complexity axis, there would be a tree that has been grown all the way to having one terminal node per observation in the data (maximum complexity).

For the complex tree, the training error can be zero (it's only non-zero if cases have different response y with all inputs x_j the same). Thus, training error is not a useful measurement of model quality and a different dataset, the test data set, is needed to assess performance. Assuming a test set is available, if for each tree size performance is measured on it, then the error curve is typically U-shaped as shown. That is, somewhere on the x-axis there is a M^*, where the test error is at its minimum, which corresponds to the optimal tree size for the given problem.

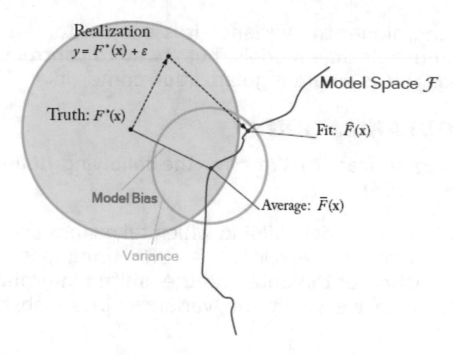

Figure 3.2: Schematic representation of Bias and Variance (adapted from (Hastie et al., 2001)).

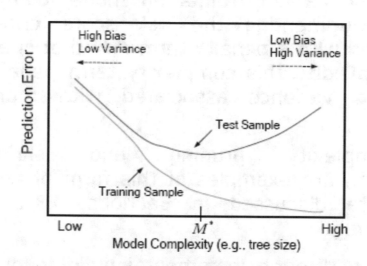

Figure 3.3: Bias-Variance trade-off as a function of model complexity (adapted from (Hastie et al., 2001)). A simple model has high bias but low variance. A complex model has low bias but high variance. To determine the optimal model a test set is required.

Finally, the bias-variance tradeoff also means that the more complex (flexible) we make the model $\hat{F}$, the lower the bias but the higher the variance it is subject to. We want to be able to use flexible models, but a way to control variance is needed. This is where regularization comes in.

3.3 REGULARIZATION

What is regularization? We offer the following definition by Rosset, S. (2003):

> "any part of model building which takes into account – implicitly or explicitly – the finiteness and imperfection of the data and the limited information in it, which we can term 'variance' in an abstract sense."

We know of at least three different ways of regularizing:

1. Explicitly via constraints on model complexity. This means augmenting the risk score criterion being minimized with a penalty term $P(F)$ that is a function of F's complexity. This complexity term penalizes for the increased variance associated with more complex models.

   Cost-complexity pruning and shrinkage-based regression are examples of this form of regularization. They are discussed in Sections 3.3.1 and 3.3.3, respectively.

2. Implicitly through incremental building of the model. This means using learning algorithms that update model parameter estimates very slowly.

   The forward stagewise linear regression algorithm is an example of this approach. It is discussed in Section 3.3.4.

3. Implicitly through the choice of robust loss functions. Since the presence of outliers is a source of variance in

the regression procedure, by using a robust loss function that is less sensitive to them, we are controlling variance, and thus doing implicit regularization.

The Huber loss (Huber, P., 1964)

$$L(y, \hat{y}) = \begin{cases} \frac{1}{2}(y - \hat{y})^2 & |y - \hat{y}| \leq \delta \\ \delta \left(|y - \hat{y}| - \delta/2 \right) & |y - \hat{y}| > \delta \end{cases}$$

is an example of a robust loss function for regression.

It is said that regularization builds on the "bet on sparsity" principle: use a method that is known to work well on "sparse" problems (where only a subset of the predictors really matter) because most methods are going to do poorly on "dense" problems.

3.3.1 REGULARIZATION AND COST-COMPLEXITY TREE PRUNING

The first example of augmenting the score criterion with a complexity term comes from the CART approach to tree pruning (Breiman et al., 1993):

$$\hat{R}_\alpha(T) = \hat{R}(T) + \alpha \cdot P(T) \tag{3.3}$$

Where

- $\hat{R}(T)$ is the error or "risk" of tree T – i.e., how good it fits the data as discussed in Chapter 2.

- $P(T) = |T|$ measures the complexity of the tree by its size – i.e., number of terminal nodes (analogous to an $L0$-norm).

- α is called the complexity parameter and represents the cost of adding another split variable to the model.

Note that the complexity term is deterministic and independent of the particular random sample; it thereby provides a stabilizing influence on the criterion being minimized. It acts as a counterbalancing force to the data-dependent part of the error; the model can get more complex if it can reduce the error by a certain amount to compensate for the increased penalty in complexity.

Our goal then is rephrased from finding the tree that has minimum risk $\hat{R}(T)$ to finding the tree that has minimum regularized risk $\hat{R}_\alpha(T)$.

The parameter α controls the degree of stabilization of the regularized component of the error. At one extreme, if α = 0, there's no regularization; a fully grown tree, T_{max}, is obtained which corresponds to the least stable estimate. At the other extreme, $\alpha \gg 0$ (much greater than 0), and the resulting tree, T_0 = root, is completely deterministic: no matter what the data indicates, the complexity penalty wins over the loss component of the risk, and a stump results. No tree can be built. In between, varying α produces a nested (finite) sequence of subtrees:

The procedure can be set up in such a way that every tree is a subset of the previous tree (Breiman et al., 1993). This "nestedness" property for trees allows the process of minimizing $R_\alpha(T)$ to work very efficiently.

Since there is a tree associated with each particular α, choosing the optimal tree is equivalent to choosing the optimal value of α, of finding α that minimizes (prediction) risk $R(T_\alpha)$.

3.3.2 CROSS-VALIDATION

As mentioned in Section 3.2, while discussing the bias-variance tradeoff, the error on the training set is not a useful estimator of model performance. What is needed is a way to estimate prediction risk, also called test risk or future risk. If not enough data is available to partition it into separate training and test sets, one may use the powerful general technique of cross-validation.

To perform what is called 3-fold cross-validation: simply, randomly split the data D into three non-overlapping groups D_1, D_2, and D_3, and generate data "folds" (partitions) with one D_i designated for testing and the other two for training (see Figure 3.4). For each fold, a model (e.g., a tree) is built on the training part of it and evaluated on the test part of it.

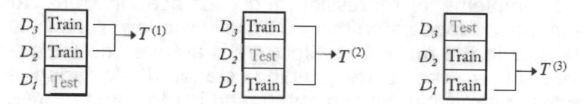

Figure 3.4: Illustration of 3-fold cross-validation. The original data set D randomly split into three non-overlapping groups D_1, D_2, and D_3. Two of the groups are allocated for training and one for testing. The process is repeated three times.

Note that every observation x_i in D is assigned to a test sub-group only once, so an indexing function can be defined,

$$v(i) : \{1, \ldots, N\} \mapsto \{1, \ldots, V\}$$

which maps the observation number, 1, ..., N, to a fold number 1, ..., V. Thus, function $v(i)$ indicates the partition in which observation i is a test observation. The cross-validated estimate of risk is then computed as:

$$\hat{R}^{CV} = \frac{1}{N} \sum_{i=1}^{N} L\left(y_i, T^{v(i)}(x_i)\right)$$

This estimate of prediction risk can be plotted against model complexity. Since varying the value of the regularization parameter α varies the complexity of the model – e.g., the size of the tree, the risk estimate can be written as a function of α:

$$\hat{R}^{CV}(\alpha) = \frac{1}{N} \sum_{i=1}^{N} L\left(y_i, T_{\alpha}^{v(i)}(x_i)\right)$$

Figure 3.5 shows a plot of $\hat{R}^{CV}(\alpha)$; it was generated using the *plotcp* command from the *rpart* package in *R* (rpart), which implements regression and classification trees. Risk estimate $\hat{R}^{CV}$ appears on the y-axis, α in the lower x-axis and tree size in the top x-axis. Since $\hat{R}^{CV}$ is an average for every value of α, the corresponding standard deviation (or standard-error) can be computed, and is represented by the short vertical bars.

To choose the optimal tree size, simply find the minimum in the $\hat{R}^{CV}(\alpha)$ plot. For the example of Figure 3.5, this minimum occurs for a tree of size 28. Sometimes the following more conservative selection approach is used: locate the minimum of the risk curve, from this point go up one standard-error, and then move horizontally left until crossing the risk curve again. The tree size value corresponding to where the crossing occurred is selected as the optimal tree size. In Figure 3.5, this occurs for a tree of size 15. This procedure, known as the 1-SE rule, corresponds to the notion that, statistically, one cannot really differentiate between those two points, so one takes the more conservative answer, which is the smaller of those two trees. This was suggested by the CART authors to protect against overfit; in our opinion, it is often over conservative.

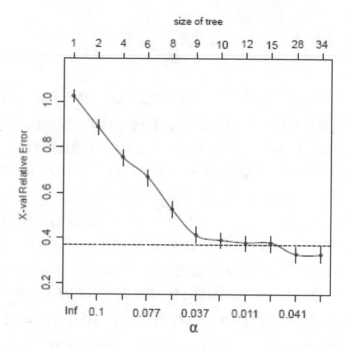

Figure 3.5: Sample plot of the cross-validated estimate of prediction risk $\hat{R}^{cv}$(error) as a function of regularization parameter *a* and model size. The optimal tree size is indicated by the minimum in the plot.

In summary, cross-validation combines (averages) measures of fit (prediction error) to correct for the optimistic nature of training error and derive a more accurate estimate of prediction risk. In Chapter 5, we'll see that ensembles are also doing a form of "averaging."

3.3.3 REGULARIZATION VIA SHRINKAGE

We now turn to regularization via shrinkage in the context of linear regression. As discussed in Chapter 2, a linear model has the form:

$$F(\mathbf{x}) = a_0 + \sum_{j=1}^{n} a_j x_j$$

and the coefficient estimation problem is simply stated as:

$$\{\hat{a}_j\} = \underset{\{a_j\}}{\arg\min} \sum_{i=1}^{N} L\left(y_i, \ a_0 + \sum_{j=1}^{n} a_j x_{ij}\right) \qquad (3.4)$$

Here, we are trying to find values for the coefficients a_j that minimize risk (as estimated by the average loss). There are two main reasons why the ordinary linear regression (OLR) solution is often not satisfactory for solving this problem. One is that often there is a high variance in the coefficient estimates (Tibshirani, R., 1996). The other is interpretation: when the number of variables is large, we would like, if possible, to identify the subset of the variables that capture the stronger effects. A solution vector a with $a_j \neq 0$ for these small subset of important variables, and $a_j = 0$ for all other variables, is termed a "sparse" solution, and it is often preferred over a "dense" solution where all coefficients are non-zero.

There are generally two types of techniques to improve OLR. One is *subset selection* (Hastie et al., 2001), which tries to identify the best subset among variables x_j to include in the model. Like the tree growing algorithm of Section 2.1, subset selection is a greedy discrete process (a variable is either in or out of the model) where often different data sub-samples give rise to different variable subsets.

The second approach is to use "shrinkage," which is a continuous process. As in Equation (3.3), shrinkage works by augmenting the error criterion being minimized with a data-independent penalty term:

$$\{\hat{a}_j\} = \underset{\{a_j\}}{\arg\min} \sum_{i=1}^{N} L\left(y_i, a_0 + \sum_{j=1}^{n} a_j x_{ij}\right) + \lambda \cdot P(a)$$

where λ controls the amount of regularization (as α did in the case of trees). The penalty function $P(\mathbf{a})$ can take different forms – e.g.,

- Ridge: $P(\mathbf{a}) = \sum_{j=1}^{n} a_j^2$

- Lasso (Tibshirani, R., 1996): $P(\mathbf{a}) = \sum_{j=1}^{n} |a_j|$

- Elastic Net (Zou and Hastie, 2005): $P_\alpha(\mathbf{a}) = \sum_{j=1}^{n} \left[\frac{1}{2}(1-\alpha) \cdot a_j^2 + \alpha \cdot |a_j| \right]$ which is a weighted mixture of the Lasso and Ridge penalties.

Note that the Ridge and Lasso penalties correspond to the $L2$- and $L1$- norm of the coefficient vector **a**, respectively. If we were to use the $L0$-norm – simply counting the number of non-zero coefficients – the penalty would be analogous to what we used in trees. And, as in the case of trees, the penalty term promotes reduced variance of the estimated values, by encouraging less complex models – i.e., those with fewer or smaller coefficients.

The Lasso differs from the Ridge penalty in that it encourages sparse coefficient vectors where some entries are set to zero. This is often the case in the presence of correlated variables; Ridge will shrink the coefficients of the correlated subset towards zero, whereas Lasso will often "select" a variable from among them. Thus, the Lasso can be viewed as a continuous version of variable selection.

Obtaining the Lasso coefficients with the squared-error loss involves solving a quadratic programming problem with constraints, which can be computationally demanding for large problems. In the past few years, however, fast iterative algorithms have been devised to solve this problem more quickly, and allow other loss functions (see Section 3.3.3).

As with α for cost-complexity tree pruning, λ is a meta-parameter of the minimization procedure that needs to be estimated (generally via cross-validation). For every value of λ, there is a corresponding "optimal" coefficient vector $\mathbf{a}(\lambda)$; having $\lambda \gg 0$ gives maximum regularization and only the constant model $y = a_0$ is produced (maximum bias and minimum variance). At the other extreme, setting $\lambda = 0$

results in no regularization (minimum bias, maximum variance or least stable estimates). This is depicted in a plot of error against the values of the regularization parameter (see Figure 3.6).

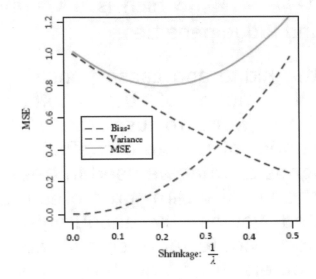

Figure 3.6: Sample plot of the estimate of prediction error (MSE) as a function of regularization parameter λ. The optimal model, $\mathbf{a}(\lambda^*)$, is found by selecting λ where the error curve has a minimum. As the amount of shrinkage increases, variance decreases and bias increases.

As discussed in Section 3.2, bias and variance are two opposing components of the error. The regularized model will have higher bias than the un-regularized one, but also smaller variance. If the decrease in variance is larger than the increase in bias, then the regularized model will be more accurate. This is illustrated in Figure 3.6 at the point where the value of shrinkage is around 0.2. With Lasso-like penalties, if the resulting solution $\mathbf{a}(\lambda^*)$ also has many coefficients $a_j = 0$, then the model will be easier to interpret.

To visualize the effect of the Lasso penalty, consider a linear problem with just two variables. Three coefficients need to be estimated: the offset, a_1, and a_2. Yet generally, the offset is not included in the penalty. From Equation (3.4), using the least-squares loss, we need to solve:

$$\{\hat{a}_0, \hat{a}_1, \hat{a}_2\} = \underset{\{a_j\}}{\arg\min} \sum_{i=1}^{N} (y_i - a_0 - a_1 x_{i1} - a_2 x_{i2})^2 + \lambda\,(|a_1| + |a_2|) \qquad (3.5)$$

Since we are using least-squares, the error function has a bowl shape with elliptical contours. We visualize the error surface as sitting on top of the plane spanned by a_1 and a_2 (see Figure 3.7).

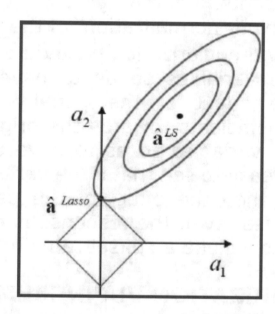

Figure 3.7: Illustration of Lasso penalty in 2D space.

The global minimum is marked by $\hat{\mathbf{a}}^{LS}$. Since it can be shown that the penalized formulation of our minimization problem, Equation (3.5), is equivalent to a constrained formulation,

$$\{\hat{a}_0, \hat{a}_1, \hat{a}_2\} = \underset{\{a_j\}}{\arg\min} \sum_{i=1}^{N} (y_i - a_0 - a_1 x_{i1} - a_2 x_{i2})^2 \quad \text{subject to} \quad (|a_1| + |a_2|) \leq \lambda$$

the complexity penalty can be represented by a diamond around the origin – i.e., the set of value pairs (a_1, a_2) for which the inequality condition $|a_1| + |a_2| \leq \lambda$ is true. If the coefficient estimates stay within the diamond, the constraint is met. Thus, the Lasso solution is the first point $\mathbf{a} = (a_1, a_2)$ where the blue contours touch the red constraint region.

In high dimensional space, this diamond becomes a rhomboid with many corners. And so if the un-regularized error surface touches one of the corners, a solution vector $\mathbf{a}$ with many entries equal to zero can be obtained. In the case of a Ridge penalty, the constrained region is actually a circle, which doesn't have corners, so one almost never gets zero value coefficients.

Finally, a note on "normalization" of the variables x_j prior to using the above regularization procedure. "Centering," or transforming the variables so as to have zero mean, is required as the effect of the penalty is to pull the corresponding coefficients towards the origin. "Scaling," or transforming the variables so as to have unit variance, is optional, but it's easy to see that if the variables have vastly different scales, then the effect of the penalty would be uneven. Sometimes, even the response is centered so that there's no need to include an offset term in the model.

3.3.4 REGULARIZATION VIA INCREMENTAL MODEL BUILDING

As the number of variables increases, the quadratic programming approach to finding the Lasso solution $\hat{\mathbf{a}}^{Lasso}$ becomes more demanding computationally. Thus, iterative algorithms producing solutions that closely approximate the effect of the lasso have been proposed. One such algorithm is called the Forward Stagewise Linear Regression (or Epsilon Stagewise Linear Regression) (Hastie et al., 2001). Figure 3.8 sketches out the algorithm.

Initialize $\hat{a}_j = 0$; $\varepsilon > 0$; M large

For $m = 1$ to M {

    // Select predictor that best fits current residuals

$$\left(\alpha^*, j^*\right) = \arg\min_{\{\alpha, j\}} \sum_{i=1}^{N} \left[y_i - \sum_{l=1}^{n} \hat{a}_l x_{il} - \alpha \cdot x_{ij} \right]^2$$

    // Increment magnitude of a_{j^*} by infinitesimal amount

$$a_{j^*} \leftarrow a_{j^*} + \varepsilon \cdot \text{sign}(\alpha^*)$$

}

write $\hat{F}(x) = \sum_{j=1}^{n} \hat{a}_j x_j$

Figure 3.8: The Epsilon Stagewise Linear Regression algorithm that approximates the Lasso solution.

The algorithm starts with all the coefficients set to zero and a small epsilon defined. Inside the loop, the algorithm first selects the predictor variable that best fits the current residuals – i.e., the difference between the response and the current model $r_i = (y_i - \sum_{l=1}^{n} \hat{a}_l x_{il})^2$. The coefficient associated with that particular variable is incremented by a small amount. The process then continues, slowly incrementing the value of one coefficient at a time for up to M iterations. M here is a meta-parameter that needs to be estimated via cross-validation. In the end, some coefficients may never get updated, staying equal to zero, and in general, $|\hat{a}_j(M)|$ is smaller than $|\hat{a}_j^{LS}|$, the least-squares estimate.

Note that M acts as the inverse of the regularization parameter λ. $M = 0$ corresponds to $\lambda >> 0$ (full regularization) and $M >> 0$ corresponds to $\lambda = 0$ (no regularization).

Another algorithm for iteratively computing the Lasso solution is LARS (Efron et al., 2004). LARS, however, is limited to the least-squares loss. The Path Seeker (PS) algorithm of Friedman (Friedman and Popescu, 2004; Friedman, J., 2008) allows the use of other differentiable loss functions – e.g., the Huber Loss. This is very desirable because least-squares is not robust in the presence of outliers.

The evolution of each coefficient from $a_j(M = 0) = 0$ to $a_j(M \gg 0) = a_j^{LS}$, as the algorithm evolves, gradually relaxing the amount of regularization, gives rise to what are called coefficient "paths" (see Figure 3.9). At the beginning, all the coefficients are equal to zero; the horizontal axis corresponds to the "shrinkage" factor $s \approx 1/M$ and the vertical axis corresponds to the values of the coefficients. As the amount of regularization is reduced, a new variable enters the model and its coefficient starts growing. The red bar indicates the optimal value of the regularization parameter, estimated via cross validation. At this point in this example, the coefficient vector **a** has five, out of eight, non-zero coefficients.

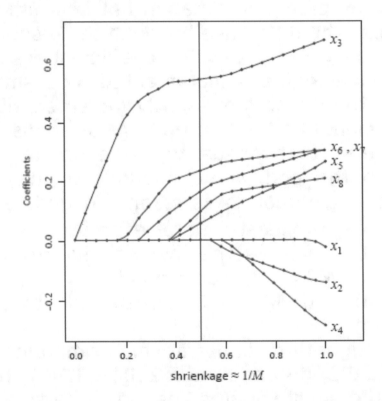

Figure 3.9: Coefficient "paths" example in a problem with eight predictor variables. With maximum regularization, $M = 0$, all coefficients $a_j = 0$. As the amount of regularization is decreased, the coefficients episodically become non-zero and gradually drift further away from zero.

3.3.5 EXAMPLE

The data for this example comes from a small study of 97 men with prostate cancer; it is available in *R*'s *faraway* package (*farway*). There are 97 observations and 9 variables (see Table 3.1). The goal is to build a linear model for predicting 'lpsa' as a function of the other eight predictors.

Table 3.1: Variables in the prostate cancer data set.

	Variable Name	Definition
1	lcavol	log(cancer volume)
2	lweight	log(prostate weight)
3	age	age
4	lbph	log(benign prostatic hyperplasia amount)
5	svi	seminal vesicle invasion
6	lcp	log(capsular penetration)
7	gleason	Gleason score
8	pgg45	percentage Gleason scores 4 or 5
9	lpsa	log(prostate specific antigen)

In *R*, accessing the data is accomplished with the following commands:

```
library(faraway)
data(prostate)
attach(prostate)
```

To check the data size and column names we use:

```
dim(prostate)
[1] 97 9
```

```
names(prostate)
[1] "lcavol"  "lweight" "age"     "lbph"    "svi"     "lcp"     "gleason"
[8] "pgg45"   "lpsa"
```

Following the analysis of this data done in (Hastie et al., 2001), we randomly split the dataset into a training set of size 67 and a test set of size 30:

```
set.seed(321); i.train <- sample(1:97, 67)
x.train <- prostate[i.train, 1:8]; y.train <- prostate[i.train, 9]
x.test <- prostate[-i.train, 1:8]; y.test <- prostate[-i.train, 9]
```

Package *lars* (lars) contains an implementation of the Lasso fitting algorithm for the least-squares loss (e.g., Equation (3.5)):

```
library(lars)
fit.lasso <- lars(as.matrix(x.train), y.train, type="lasso")
plot(fit.lasso, breaks=F, xvar="norm")
```

The last command plots the coefficient paths for the prostate fit (see Figure 3.10(a)) as a function of the fraction $s = |\mathbf{a}| / \text{Max} |\mathbf{a}|$ (the $L1$ norm of the coefficient vector, as a fraction of the maximal $L1$ norm), which is another way of expressing the amount of regularization done. To select the optimal value of the coefficient vector $\mathbf{a}$ along these paths, we need to estimate prediction error as a function of s, via cross validation:

```
cv.lasso <- cv.lars(as.matrix(x.train), y.train, type="lasso")
```

which results in the plot of Figure 3.10(b). To select the minimum according to the 1-SE rule discussed earlier (Section 3.3.2), we use:

```
i.min <- which.min(cv.lasso$cv)
i.se <- which.min(abs(cv.lasso$cv -
        (cv.lasso$cv[i.min]+cv.lasso$cv.error[i.min])))
s.best <- cv.lasso$fraction[i.se]
```

a(*s.best*) corresponds to the most parsimonious model with a prediction error within one standard error of the minimum.

The optimal coefficient values can now be retrieved:

```
predict.lars(fit.lasso, s = s.best, type = "coefficients",
             mode = "fraction")
    lcavol   lweight       age      lbph       svi       lcp
0.5040864 0.1054586 0.0000000 0.0000000 0.3445356 0.0000000
  gleason     pgg45
0000000 0.0000000
```

We observe that only three coefficients are non-zero. The corresponding error on the test set is computed by:

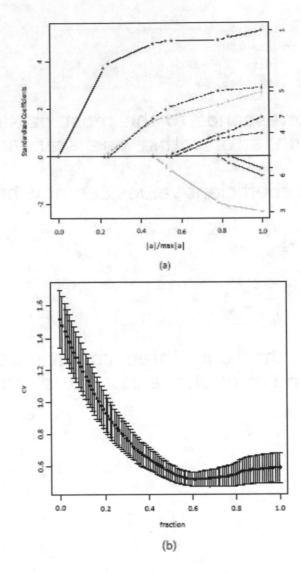

Figure 3.10: Coefficient "paths" and cross-validation error for a Lasso fit to the Prostate data.

```
y.hat.test <- predict.lars(fit.lasso, x.test, s = s.best, type = "fit",
                           mode = "fraction")
sum((y.hat.test$fit - y.test)^2) / 30
[1] 0.5775056
```

which compares favorably to an OLR fit using all coefficients:

```
fit.lm <-lm(lpsa ~ ., data = prostate[i.train,])
fit.lm$coeff
  (Intercept)         lcavol        lweight              age           lbph
-0.787006412    0.593999810     0.499004071   -0.012048469    0.093761682
          svi            lcp        gleason            pgg45
 0.847659670   -0.023149568     0.180985575   -0.002979421

y.hat.lm.test <- predict(fit.lm, prostate[-i.train,])
sum((y.hat.lm.test - prostate$lpsa[-i.train])^2) / 30
[1] 0.6356946
```

3.3.6 REGULARIZATION SUMMARY

Let's revisit Figure 3.2, which provided a schematic of what bias and variance are. The effect of adding the complexity penalty $P(\mathbf{a})$ to the risk criterion is to "restrict" the original model space (see Figure 3.10). The "distance" between the average regularized estimator $\bar{F}^{Lasso}$ and the truth F^* is larger than between the average un-regularized estimator $\bar{F}$ and the truth F^*; that is, we have increased model bias. However, the corresponding variance (spread around the average) is smaller. When the reduction in variance is larger than the increase in bias, the resulting model will be more accurate.

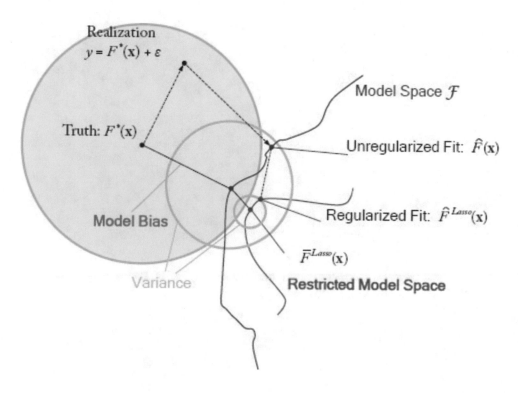

Figure 3.11: Bias-Variance schematic with effect of regularization (adapted from (Hastie et al., 2001)).

[1] Unless two cases have identical input values and different output values.

## CHAPTER 4

Importance Sampling and the Classic Ensemble Methods

In this chapter, we provide an overview of Importance Sampling Learning Ensembles (ISLE) (Friedman and Popescu, 2003). The ISLE framework allows us to view the classic ensemble methods of Bagging, Random Forest, AdaBoost, and Gradient Boosting as special cases of a single algorithm. This unified view clarifies the properties of these methods and suggests ways to improve their accuracy and speed.

The type of ensemble models we are discussing here can be described as an additive expansion of the form:

$$F(\mathbf{x}) = c_0 + \sum_{m=1}^{M} c_m T_m(\mathbf{x})$$

where the $\{T_m(\mathbf{x})\}_1^M$ are known as *basis functions* or also called *base learners*. For example, each T_m can be a decision tree. The ensemble is a linear model in a (very) high dimensional space of derived variables. Additive models like this are not new: Neural Networks (Bishop, C., 1995), Support Vector Machines (Scholkopf et al., 1999), and wavelets in Signal Processing (Coifman et al., 1992), to name just a few, have a similar functional form.

It's going to be convenient to use the notation $T(\mathbf{x}; \mathbf{p}_m)$ when referring to each base learner T_m. That is, each base learner is described by a set of parameters or parameter vector $\mathbf{p}$. For example, if T_m is a neural net, $\mathbf{p}_m$ corresponds to the weights that define the neural net. If T_m is a tree, $\mathbf{p}_m$ corresponds to the splits that define the tree. Each possible base learner can then be thought of as a "point" in a high-dimensional parameter space $\mathbf{P}$.

With this notation, the ensemble learning problem can be stated as follows: find the points $\mathbf{p}_m \in \mathbf{P}$ and the constants $c_m \in R$ (real numbers) that minimize the average loss:

$$\{\hat{c}_m, \ \hat{p}_m\}_o^M = \min_{\{c_m, \ p_m\}_o^M} \sum_{i=1}^{N} L\left(y_i, \ c_0 + \sum_{m=1}^{M} c_m T(\mathbf{x}; \ p_m)\right) \qquad (4.1)$$

Much like in the case of decision tree regression (see Section 2.2), the joint optimization of this problem is very difficult. A heuristic two-step approach is useful:

1. Choose the points $\mathbf{p}_m$. This is equivalent to saying "choose a subset of M base learners out of the space of all possible base learners from a pre-specified family" – e.g., the family of 5-terminal node trees.

2. Determine the coefficients, or weights, c_m via regularized linear regression (see Chapter 3 for an overview of regularization).

Before introducing the details of the ensemble construction algorithm, consider the example of Figure 4.1. It is a 2-input, 2-class problem that has a linear decision boundary given by the diagonal line. As discussed in Chapter 2, linear decision boundaries such as this one are a hard case for trees, which have to build a stair-like approximation. The decision boundary built by a tree ensemble based on Boosting (Section 4.5) is still piece-wise constant but with a finer resolution, thus better capturing the diagonal boundary.

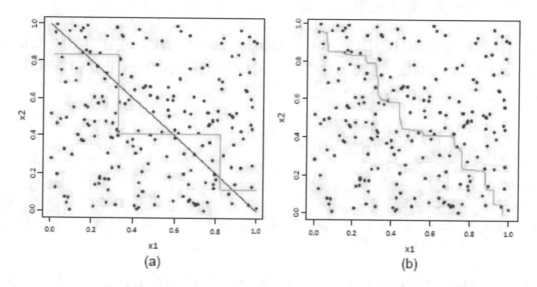

Figure 4.1: Comparison of decision boundaries generated by a single decision tree (a) and a Boosting-based ensemble model (b). The true boundary is the black diagonal. The ensemble approximation is superior.

Table 4.1 includes code for generating and plotting the data. In R, fitting a classification tree to this data is accomplished with the following commands:

```
library(rpart)
set.seed(123)

tree <- rpart(y ~ . , data = data2d, cp = 0, minsplit = 4, minbucket = 2,
            parms = list(prior=c(.5,.5)))
```

To prune the tree using the 1-SE rule, discussed in Chapter 3, we use:

```
i.min <- which.min(tree$cptable[,"xerror"])
```

```
i.se <- which.min(abs(tree$cptable[,"xerror"] -
                          (tree$cptable[i.min,"xerror"]
                          + tree$cptable[i.min,"xstd"])))
alpha.best <- tree$cptable[i.se,"CP"]
tree.p <- prune(tree, cp = alpha.best)
```

And to plot the decision boundary, we evaluate the tree on a 100 × 100 grid and color the points where the two classes are equally probable:

```
xp <- seq(0, 1, length = 100)
yp <- seq(0, 1, length = 100)
data2dT <- expand.grid(x1 = xp, x2 = yp)
Z <- predict(tree.p, data2dT)
zp.cart <- Z[,1] - Z[,2]
contour(xp, yp, matrix(zp.cart, 100), add=T, levels=0, labcex=0.9, labels="",
        col = "green", lwd=2)
```

The boosting library available in R, *gbm* (gbm), requires assigning numeric labels to the training data points – i.e., a +1 to the blue cases and a 0 to the red cases. This is easily done with:

```
y01 <- rep(0, length(data2d$y))
y01[which(data2d$y == 'blue')] <- 1
data4gbm <- data.frame(y01, x1, x2)
```

and fitting the boosting model to this data is accomplished with the following commands:

```
library(gbm)
set.seed(123)
boostm <- gbm(y01 ~ ., data = data4gbm, distribution = "bernoulli",
              n.trees = 100, interaction.depth = 2,
              n.minobsinnode = 4, shrinkage = 0.1,
              bag.fraction = 0.5, train.fraction = 1.0, cv.folds = 10)
```

here too we rely on cross-validation to choose the optimal model size:

```
best.iter <- gbm.perf(boostm, method = "cv")
```

as we did for a single tree, we plot the decision boundary by evaluating the model on the 100 × 100 grid generated above:

```
Z.gbm <- predict(boostm, data2dT, n.tree = best.iter, type = "response")
zp.gbm <- 1 - 2*Z.gbm
contour(xp, yp, matrix(zp.gbm, 100), add=T, levels=0, labcex=0.9, labels="",
        col = "green", lwd=2)
```

Table 4.1: Sample R code for generating the 2-input, 2-class data of Figure 4.1.

```
genData <- function(seed, N) {
  set.seed(seed)
  x1 <- runif(N)
  x2 <- runif(N)
  y <- rep("", N)
  for (i in 1:N) {
    if ( x1[i] + x2[i] >  1.0) {
      y[i] <- "blue"
    }
    if ( x1[i] + x2[i] <  1.0) {
      y[i] <- "red"
    }
    if ( x1[i] + x2[i] ==  1.0) {
      if ( runif(1) < 0.5 ) {
        y[i] <- "red"
      } else {
        y[i] <- "blue"
      }
    }
  }
  y <- as.factor(y)
  return(data.frame(x1, x2, y))
}
```
```
## Generate data
data2d <- genData(123, 200)
summary(data2d$y)
blue   red
 108    92

## Plot data
i.red <- y == 'red'
i.blue <- y == 'blue'
plot(x1, x2, type="n")
points(x1[i.blue], x2[i.blue], col = "blue", pch = 19)
points(x1[i.red], x2[i.red], col = "red", pch =  19)
lines(c(1,0), c(0,1), lwd=2)
```

4.1 IMPORTANCE SAMPLING

The ISLE framework helps us develop an answer to the question of how to judiciously choose the basis functions. The goal is to find "good" $\{p_m\}_1^M$ so that the ensemble-based approximation is "close" to the target function:

$$F(x; \{p_m\}_1^M, \{c_m\}_0^M) = c_0 + \sum_{m=1}^{M} c_m T(x; p_m)$$
$$\approx F^*(x) \tag{4.2}$$

ISLE makes a connection with numerical integration. The key idea is that the function we're trying to fit above is analogous to a high-dimensional integral:

$$\int_P I(p) \, \partial p \approx \sum_{m=1}^{M} w_m I(p_m) \tag{4.3}$$

Consider, for example, the problem of integrating the function $I(p)$ drawn in red in Figure 4.2. The integral is approximated by the average of the function evaluated at a set of points p_1, p_2,..., p_M. The usual algorithms uniformly choose the points p_m at which the integrand $I(p)$ is evaluated. Importance Sampling, on the other hand, recognizes that certain values of these p_m variables have more impact on the accuracy of the integral being estimated, and thus these "important" values should be emphasized by sampling more frequently from among them.

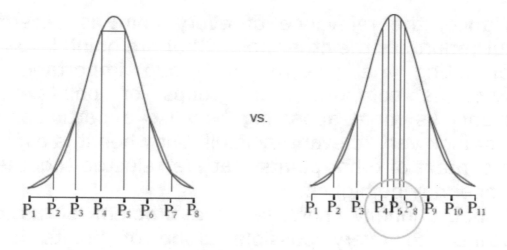

Figure 4.2: Numerical integration example. Accuracy of the integral improves when we choose more points from the circled region.

Thus, techniques such as Monte Carlo integration used for computing the approximation in Equation (4.3) can be studied for developing algorithms for finding the approximation of Equation (4.2).

4.1.1 PARAMETER IMPORTANCE MEASURE

In order to formalize the notion of choosing good points from among the space **P** of all possible $\mathbf{p}_m$, we need to define a sampling distribution. Thus, assume that it is possible to define a sampling probability density function (pdf) $r(\mathbf{p})$ according to which we are going to draw the points $\mathbf{p}_m$–i.e., $\{\mathbf{p}_m \sim r(\mathbf{p})\}_1^M$. The simplest approach would be to have $r(\mathbf{p})$ be uniform, but this wouldn't have the effect of "encouraging" the selection of important points $\mathbf{p}_m$. In our Predictive Learning problem, we want $r(\mathbf{p})$ to be inversely related to the "risk" of $\mathbf{p}_m$ (see Chapter 2 Equation (2.1)). If $T(\mathbf{x}; \mathbf{p}_m)$ has high error, then $\mathbf{p}_m$ has low relevance and $r(\mathbf{p})$ should be small.

In Monte Carlo integration, point importance can also be measured in "single" or "group" fashion. In single point importance, the relevance of every point is determined without regard for the other points that are going to be used in computing the integral. In group importance, the relevance is computed for groups of points. Group importance is more appealing because a particular point may not look very relevant by itself, but when it is evaluated in the context of other points that are selected together, its relevance may be higher.

Computationally, however, the problem of assigning importance to every possible group of points is very demanding. Thus, one often uses the "sequential approximation" to group importance, where the relevance of a particular point is judged in the context of the points that have been selected so far but not the points that are yet to be selected.

Like any density distribution, $r(\mathbf{p})$ can be characterized by its "center" and "width." Assume $\mathbf{p}*$ represents the best single base learner – e.g., the single tree $T(\mathbf{x}; \mathbf{p}*)$ that minimizes risk, and that $r(\mathbf{p})$ is centered at this $\mathbf{p}*$ (see Figure 4.3). In Figure 4.3(a), $r(\mathbf{p})$ is "narrow," which means points $\mathbf{p}_m$ will be selected from a small vicinity of $\mathbf{p}*$. The resulting ensemble $\{T_m(\mathbf{x})\}_1^M$ would be made of many "strong" base learners T_m with $\mathrm{Risk}(T(\mathbf{x}; \mathbf{p}_m)) \approx \mathrm{Risk}(T(\mathbf{x}; \mathbf{p}*))$. In this case, the predications made by the individual $T(\mathbf{x}; \mathbf{p}_m)$ are highly correlated, and the combination won't necessarily lead to a significant accuracy improvement.

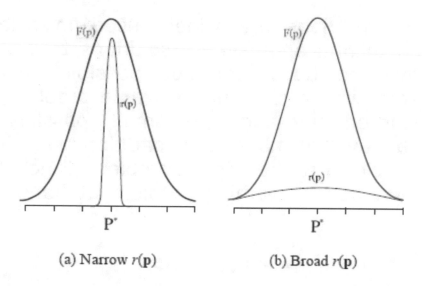

(a) Narrow $r(\mathbf{p})$ (b) Broad $r(\mathbf{p})$

Figure 4.3: Characterization of point importance measure $r(\mathbf{p})$ in terms of its width. $F(\mathbf{p})$ denotes the function being integrated and $\mathbf{p}*$ denotes the best single point.

In Figure 4.3(b), $r(\mathbf{p})$ is "broad," which means points $\mathbf{p}_m$ will be selected from a wide region in the integrand's domain. This will translate into a "diverse" ensemble where predictions from individual $T(\mathbf{x};\ \mathbf{p}_m)$ are not highly correlated with each other. However, if the final base learner set contains many weak ones with Risk $(T(\mathbf{x};\ \mathbf{p}_m)) \gg$ Risk$(T(\mathbf{x};\ \mathbf{p}*))$, then our ensemble will have poor performance.

In summary, the empirically observed relationship between ensemble performance and the strength/correlation of the constituent base learners can be understood in terms of the width of $r(\mathbf{p})$. The optimal sampling distribution depends on the unknown target function $F^*(\mathbf{x})$, and thus $r(\mathbf{p})$ is also unknown. But an approximation $\hat{r}(\mathbf{p})$ can be built based on a simulation of the sampling process.

4.1.2 PERTURBATION SAMPLING

The heuristic to simulate the process of sampling from $r(\mathbf{p})$ is based on a technique called perturbation sampling.

Perturbation methods are widely used in science and engineering to find approximate solutions to problems for which exact solutions are not possible. Perturbation methods start with a simplified solvable problem, which is then "perturbed" by adding a small, possibly random, change in a way that makes the conditions satisfied by the new solution closer to the original problem (Hinch, E., 1991).

In the Predictive Learning problem at hand, we know how to solve

$$\mathbf{p}^* = \arg\min_{\mathbf{p}} E_{\mathbf{xy}} L(y, T(\mathbf{x}; \mathbf{p})) \tag{4.4}$$

that is, we know how to find the best single base learner. There are at least three aspects of this problem that can be perturbed:

1. Perturbation of the data distribution $< \mathbf{x}, y >$. For instance, by re-weighting the observations.

2. Perturbation of the loss function $L(\cdot)$. For example, by modifying its argument.

3. Perturbation of the search algorithm used to find $\min_{\mathbf{p}}$.

Repeatedly finding the solution to a particular perturbed version of Equation (4.4) is then equivalent to sampling $\mathbf{p}_m$'s according to $\hat{r}(\mathbf{p})$, with the width of $\hat{r}(\mathbf{p})$ controlled by the degree of perturbation done.

In terms of perturbation sampling, generating the ensemble members $\{\mathbf{p}_m\}_1^M$ is thus expressed with the following algorithm:

$$\text{For } m = 1 \text{ to } M \ \{$$
$$\mathbf{p}_m = \text{PERTURB}_m \left\{ \arg\min_{\mathbf{p}} E_{\mathbf{xy}} L\left(y, T(\mathbf{x}; \mathbf{p})\right) \right\}$$
$$\}$$

where PERTURB{.} is a small perturbation of the type mentioned above. Later in this chapter, we will see examples of all these types of perturbations. For instance, the AdaBoost algorithm reweights the observations (Section 4.5), Gradient Boosting modifies the argument to the loss function (Section 4.6), and Random Forest (Section 4.4) modifies the search algorithm.

4.2 GENERIC ENSEMBLE GENERATION

The generic ensemble generation algorithm described above was formalized by Friedman and Popescu (2003) and consists of two steps: select points and fit coefficients. In detail:

· Step 1 - Choose $\{\mathbf{p}_m\}_1^M$:

$$F_0(\mathbf{x}) = 0 \qquad \text{Line 1}$$
$$\text{For } m = 1 \text{ to } M \; \{ \qquad \text{Line 2}$$
$$\mathbf{p}_m = \arg\min_{\mathbf{p}} \sum_{i \in S_m(\eta)} L\big(y_i, F_{m-1}(\mathbf{x}_i) + T(\mathbf{x}_i; \mathbf{p})\big) \qquad \text{Line 3}$$
$$T_m(\mathbf{x}) = T(\mathbf{x}; \mathbf{p}_m) \qquad \text{Line 4}$$
$$F_m(\mathbf{x}) = F_{m-1}(\mathbf{x}) + \upsilon\, T_m(\mathbf{x}) \qquad \text{Line 5}$$
$$\}$$
$$\text{write } \{T_m(\mathbf{x})\}_1^M \qquad \text{Line 6}$$

Algorithm 4.1

The algorithm starts the ensemble F_0 with some constant function (Line 1); it could be zero or another suitable constant. Then, at each iteration, a new base-learner T_m is added into the collection (Line 3). F_{m-1} represents the ensemble of base learners up to step $m - 1$.

The expression $\mathbf{p}_m = \arg\min_{\mathbf{p}} \cdots$ in Line 3 is a slightly modified version of Equation (4.4), which stands for finding the best (lowest error) base-learner on the selected data; here the inclusion of F_{m-1} as an argument to the loss function corresponds to the implementation of the "sequential approximation" to group importance. That is, we want to find the base-learner that in combination with the ones that have already been selected best approximates the response. The ensemble is then updated with the newly selected base-learner T_m (Line 5). After M base learners have been built, the algorithm terminates in Line 6. Notice the similarity with the forward stagewise fitting procedure for regularized regression of Chapter 3.

Three parameters control the operation of the algorithm, L, η, υ:

- L: the loss function. For example, the AdaBoost algorithm, discussed in Section 4.5, uses the exponential loss.

- η: controls the amount of perturbation to the data distribution. The notation $S(\eta)$ in Line 3 indicates a random sub-sample of size $\eta \leq N$, less than or equal to the original data size.

  Intuitively, smaller values of η will increase the ensemble diversity. As the size of the sample used to fit each base-learner is reduced, the perturbation is larger. η also has an impact on computing time because, if every T_m is built on 10 percent of the data, for example, the total time to built the ensemble can be reduced substantially.

- υ: controls the amount of perturbation to the loss function. Note that Line 5 in the algorithm can be alternatively expressed as $F_{m-1}(\mathbf{x}) = \upsilon \sum_{k=1}^{m-1} T_k(\mathbf{x})$, and thus υ controls how much the approximation built up to the current iteration influences the selection of the next base learner. (F_{m-1} is sometimes referred to as the "memory" function). Accordingly, having $\upsilon = 0$ is equivalent to using "single point" importance, and having $0 < \upsilon \leq 1$ corresponds to the sequential approximation to "group" importance.

· Step 2: Choose coefficients $\{c_m\}_0^M$

Once all the base learners $\{T_m(\mathbf{x})\}_1^M = \{T(\mathbf{x}; \mathbf{p}_m)\}_1^M$ have been selected, the coefficients are obtained by a regularized linear regression:

$$\{\hat{c}_m\} = \underset{\{c_m\}}{\arg\min} \sum_{i=1}^{N} L\left(y_i, c_0 + \sum_{m=1}^{M} c_m T_m(x_i)\right) + \lambda \cdot P(c) \qquad (4.5)$$

where $P(\mathbf{c})$ is the complexity penalty and λ is the meta-parameter controlling the amount of regularization as discussed in Chapter 3.

Regularization here helps reduce bias, in addition to variance, because it allows the use of a wider sampling distribution $\hat{r}_{(\mathbf{p})}$ in Step 1. A wide sampling distribution permits many weak learners to be included in the ensemble. If too many weak learners are included, however, the performance of the ensemble might be poor because those weak learners are fitted to small fractions of the data and could bring noise into the model. But using a narrow distribution isn't desirable either as it would lead to an ensemble where the predications made by the individual T ($\mathbf{x}$; $\mathbf{p}_m$) are highly correlated. With a regularization-based post-processing step, a wide distribution can be used in Step 1, and then using a Lasso-like penalty in Step 2, the coefficients c_m of those base learners that are not very helpful can be forced to be zero.

This regularization step is done on an $N \times M$ data matrix, with the output of the T_m's as the predictor variables. Since M can be on the order of tens of thousands, using traditional algorithms to solve it can be computationally prohibitive. However, fast iterative algorithms are now available for solving this problem for a variety of loss functions, including "GLMs via Coordinate Descent" (Friedman et al., 2008) and "Generalized Path Seeker" (Friedman and Bogdan, 2008).

We now turn our attention to the classic ensemble methods of Bagging, Random Forest, AdaBoot, and Gradient Boosting. To view them as special cases of the Generic Ensemble Generation procedure, we will copy Algorithm 4.1. and explain how the control parameters are set in each case. We will also discuss the improvements that the ISLE framework suggests in each case.

4.3 BAGGING

Breiman's Bagging, short for Bootstrap Aggregation, is one of the earliest and simplest ensembling algorithms (Breiman, L., 1996). In Bagging, the T_m's are fitted to bootstrap replicates of the training set D (a bootstrap replicate D' is obtained by sampling randomly from D with replacement – i.e., an observation $\mathbf{x}_i$ may appear multiple times in the sample). After the base learners have been fit, the aggregated response is the average over the T_m's when predicting a numerical outcome (regression), and a plurality vote when predicting a categorical outcome (classification). Table 4.2 shows Bagging expressed as an ISLE algorithm.

Table 4.2: Bagging as an ISLE-based algorithm.

- $L(y, \hat{y}) = (y - \hat{y})^2$

- $\upsilon = 0$

- $\eta = N/2$

- $T_m(\mathbf{x})$: large, un-pruned trees

- $c_o = 0, \ \{c_m = 1/M\}_1^M$

$F_0(\mathbf{x}) = 0$

For $m = 1$ to M {

$\quad \mathbf{p}_m = \arg\min_{\mathbf{p}} \sum_{i \in S_m(\eta)} L(y_i, F_{m-1}(\mathbf{x}_i) + T(\mathbf{x}_i; \mathbf{p}))$

$\quad T_m(\mathbf{x}) = T(\mathbf{x}; \mathbf{p}_m)$

$\quad F_m(\mathbf{x}) = F_{m-1}(\mathbf{x}) + \upsilon \cdot T_m(\mathbf{x})$

}

write $\{T_m(\mathbf{x})\}_1^M$

$\upsilon = 0$ means there is no "memory," and each T_m is fitted independently without any awareness of the other base-learners that have already been selected. $\eta = N/2$ means each T_m is built using half-size samples without replacement (it can be shown that using size $N/2$ samples without replacement is roughly equivalent to using size N samples with replacement (Friedman and Popescu, 2003), as called for in the original formulation of Bagging). The base learners T_m were typically large un-pruned trees. We'll see in Section 4.7, devoted to the MART algorithm, that there are situations where this is not the best thing to do.

The coefficients in Bagging are $c_o = 0$, $\{c_m = 1/M\}_1^M$, namely, a simple average. They are not fitted to the data. This averaging for regression is justifiable assuming a squared-error loss. Although the original formulation called for a majority vote to be used for classification problems, averaging would also work since trees already estimate class probabilities.

Bagging does not fit the coefficients to the data after the trees have been selected. It just simply assigns the same coefficient to all the trees. In summary, Bagging perturbs only one of the three possible "knobs," perturbing the data distribution only.

Note that because Bagging has no memory, it is easily parallelizable: the task of building each T_m can be spun to a different CPU (Panda et al., 2009). This divisibility is a computational advantage of the algorithm.

What improvements can be made to Bagging when viewed from the generic ensemble generation algorithm perspective? We suggest:

1. Use a different sampling strategy. There is no theoretical reason to prefer $\eta = N/2$ over other possible values. Using smaller values, e.g., < 50% samples, will make the ensemble build faster.

2. Fit the coefficients to the data. That is, instead of using a simple average, $\{c_m = 1/M\}_1^M$, the coefficients can be obtained by a regularized linear regression (see Equation (4.5)).

In R, an implementation of Bagging is available in the *ipred* package (ipred).

Figure 4.4 and Figure 4.5 (from Elder and Ridgeway (1999)) show how the out-of-sample error for a suite of regression and classification problems, respectively, reduced on all of the problems attempted, when going from a single tree to a bagged ensemble model. Though these are primarily academic datasets, our practical experience has been consistent with this finding; so far, we have always seen ensembles of trees beat single trees in out-of-sample performance. (This is not the case for ensembles vs. single models, in general, as can be seen from the examples in Chapter 1, but just for ensembles of decision trees vs. single trees.)

4.3.1 EXAMPLE

We now look at an example of Bagging in action using a very simple data set borrowed from the *Elements of Statistical Learning* book (Hastie et al., 2001). (Which is a must-have reference in our field.)

There are five predictor variables, each having a "standard" Normal distribution with pairwise correlation of 0.95. The response variable is created so as to generate a 2-class problem according to the rule

$$P(y = 1|x_1 \le 0.5) = 0.2, \quad P(y = 1|x_1 > 0.5) = 0.8$$

Thus, the minimum possible error rate (known as the Bayes error rate) is 0.2. Table 4.3 has R code for generating the data.

It follows from its definition that the response depends on x_1 only. A small training data set is generated with $N = 30$ (see functions *genPredictors* and *genTarget* in Table 4.3), from which 200 bootstrap samples are created (see function *genBStrapSamp*). A separate test set of size 2000 is also generated to evaluate an ensemble model built using Bagging. Table 4.4 shows R code for building the ensemble using these bootstrap replicates of the training data set.

The first two trees are shown in Figure 4.6. In the first, we see that the root split is on x_1, and that the split definition closely matches the rule used to generate the data. In the second tree, the first split is on x_2; because the variables are highly correlated and the sample size small, it is possible for x_2 to take the role of x_1. Bagging then continues along these lines, fitting a tree for each bootstrap sample.

Figure 4.7 shows the test error of the bagged ensemble. The Bayes error (minimum possible, with unlimited data) is indicated with a green line. The error of a single tree, fitted to the original training set, is indicated with a red line. This single tree happens to be a "stump" after cross-validation was used to decide the optimal pruning. Selecting only a stump isn't surprising given the high variance on this data due to the correlation in the predictors.

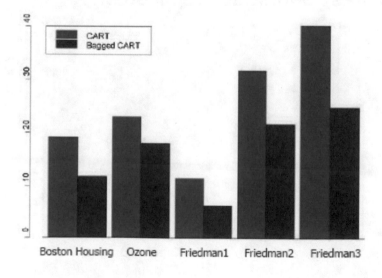

Figure 4.4: Bagged trees better than single tree for five regression problems taken from the UC Irvine data repository.

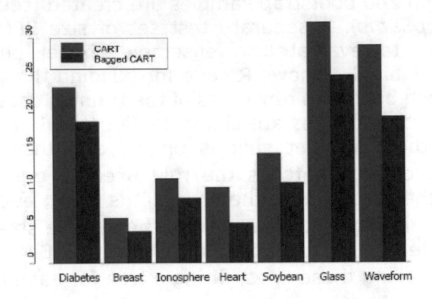

Figure 4.5: Bagged trees better than single tree on seven classification problems taken from the UC Irvine data repository.

Table 4.3: Sample R code for generating artificial data. Function *genPredictors* generates 5-dimensional observations, function *genTarget* generates the corresponding response variable and function *genBStrapSamp* generates "bootstrap" replicates.

```r
genPredictors <- function(seed = 123, N = 30) {
  # Load package with random number generation
  # for the multivariate normal distribution
  library(mnormt)
  # 5 "features" each having a "standard" Normal
  # distribution with pairwise correlation 0.95
  Rho <- matrix(c(1,.95,.95,.95,.95,
            + .95, 1,.95,.95,.95,
            + .95,.95,1,.95,.95,
            + .95,.95,.95,1,.95,
            + .95,.95,.95,.95,1), 5, 5)
  mu <- c(rep(0,5))
  set.seed(seed);
  x <- rmnorm(N, mu, Rho)
  colnames(x) <- c("x1", "x2", "x3", "x4", "x5")
  return(x)
}
```

```r
genTarget <- function(x, N, seed = 123) {
# Response Y is generated according to:
#    Pr(Y = 1 | x1 <= 0.5) = 0.2,
#    Pr(Y = 1 | x1 > 0.5) = 0.8
  y <- c(rep(-1, N))
  set.seed(seed);
  for (i in 1:N) {
    if ( x[i,1] <= 0.5 ) {
      if ( runif(1) <= 0.2 ) {
         y[i] <- 1
      } else {
         y[i] <- 0
      }
    } else {
      if ( runif(1) <= 0.8 ) {
         y[i] <- 1
      } else {
         y[i] <- 0
      }
    }
  }
  return(y)
}
```

```r
genBStrapSamp <- function(seed = 123, N = 200, Size = 30) {
  set.seed(seed)
  sampleList <- vector(mode = "list", length = N)
  for (i in 1:N) {
    sampleList[[i]] <- sample(1:Size, replace=TRUE)
  }
  return(sampleList)
}
```

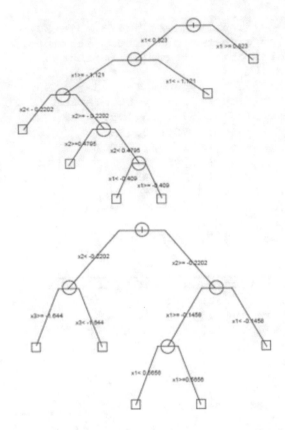

Figure 4.6: First two classification trees in a bagged ensemble for the problem of Table 4.3.

Table 4.4: Sample R code for generating a classification ensemble based on the Bagging algorithm. Function *fitClassTree* fits a single decision tree to a given training data set. Function *fitBStrapTrees* keeps track of the trees built for each "bootstrap" replicate.

```
fitBStrapTrees <- function(data, sampleList, N) {
  treeList <- vector(mode = "list", length = N)
  tree.params=list(minsplit = 4, minbucket = 2, maxdepth = 7)
  for (i in 1:N) {
    treeList[[i]] <- fitClassTree(data[sampleList[[i]],],
                                  tree.params)
  }
  return(treeList)
}

fitClassTree <- function(x, params, w = NULL,
                         seed = 123) {
  library(rpart)
  set.seed(seed)
  tree <- rpart(y ~ ., method = "class",
                data = x, weights = w, cp = 0,
                minsplit = params.minsplit,
                minbucket = params.minbucket,
                maxdepth = params.maxdepth)
  return(tree)
}
```

The error of the Bagged ensemble is shown (in blue) as a function of the ensemble size. As the number of trees increases, the error generally decreases until it eventually flattens out. Bagging effectively succeeds in smoothing out the variance here and hence reduces test error.

The flattening of the error of the bagged ensemble is not unique behavior for this example. As discussed in the next section, Bagging for regression won't "over-fit" the data when the number of trees is arbitrarily increased.

4.3.2 WHY IT HELPS?

So why does bagging help? The main reason is that Bagging reduces the variance and leaves bias unchanged. Under the squared-error loss, $L(y, \hat{y}) = (y - \hat{y})^2$, the following formal analysis can be done. Consider the "idealized" bagging (aggregate) estimator $\bar{F}(x) = E\hat{F}_Z(x)$ – i.e., the average of the $\hat{F}_Z$'s, each fit to a bootstrap data set $Z = \{y_i, x_i\}_1^N$. For this analysis, these Z's are sampled from the actual population distribution (not the training data). Using simple linear algebra and properties of the expectation operator, we can write:

$$E\left[Y - \hat{F}_Z(x)\right]^2 = E\left[Y - \bar{F}(x) + \bar{F}(x) - \hat{F}_Z(x)\right]^2$$
$$= E\left[Y - \bar{F}(x)\right]^2 + E\left[\hat{F}_Z(x) - \bar{F}(x)\right]^2$$
$$\geq E\left[Y - \bar{F}(x)\right]^2$$

The above inequality establishes that the error of one single $\hat{F}_Z$ is always greater than the error of $\bar{F}$. So true population aggregation never increases mean square error, and it often reduces it.

The above argument does not hold for classification because of the "non-additivity" of bias and variance. That is, bagging a bad classifier can make it worse!

4.4 RANDOM FOREST

The Random Forest technique (Breiman, L., 2001) is Bagging plus a "perturbation" of the algorithm used to fit the base learners. The specific form of the perturbation is called "subset splitting."

As discussed in Section 2.1, the process of building a single tree entails successively finding the best split at each node by considering every possible variable in turn; for each variable, every possible split point is evaluated. In Random Forest, only a random subset of the variables is considered at each node. Breiman's recommended size for the random subset is $n_s = \lfloor \log_2(n) + 1 \rfloor$. Thus, with 100 predictor variables; every time that a tree node needs to be split, a random sample of 11 predictors is drawn.

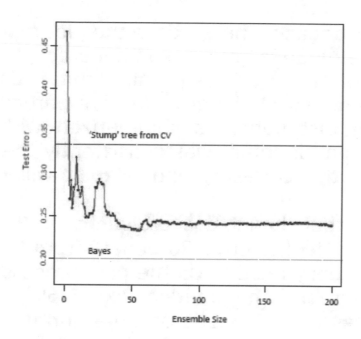

Figure 4.7: Evaluation error rate for a bagged ensemble as a function of the number of trees. The ensemble was fit to the data of Table 4.2.

Clearly, Random Forest falls under the perturbation sampling paradigm we have discussed. And, compared to Bagging, its $\{T_m(x)\}_1^M$ set has increased diversity – i.e., wider $\hat{r}(p)$, the width of which is (inversely) controlled by n_s. Random Forest also has a significant speed improvement advantage over Bagging since fewer splits need to be evaluated at each node.

As in the case of Bagging, two potential improvements are possible: 1. use of a different data sampling strategy (not fixed to bootstrap samples), and 2. fit the quadrature coefficients to the data.

In R, Random Forest for classification and regression is available in the *randomForest* package (randomForest).

How do Bagging and Random Forest compare in terms of accuracy? Figure 4.8 shows the results of a simulation study with 100 different target functions conducted by Friedman and Popescu (2004). In the x-axis, we have four algorithms: Bagging (Bag), Random Forest (RF), and one improvement of each. The y-axis in the plot represents what is called "comparative" root mean squared (RMS) error: for every problem, the error of each algorithm is divided by the error of the single best algorithm for that particular problem; the resulting distribution is summarized with a boxplot. Thus, if a given algorithm was consistently the best across all problems, the corresponding boxplot will only show a horizontal line at 1.0.

One sees that the increased diversity in Random Forest often results in higher error. To benefit from that additional diversity, we really need to do the post-processing phase of the ISLE framework. The notation "xxx_6_5%_P" indicates 6 terminal nodes trees (instead of large unpruned ones), 5% samples without replacement (instead of bootstrap ones), and Post-processing – i.e., fitting the $\{c_m\}_0^M$ coefficients to the data using regularized regression.

In both cases, the ISLE-based versions of Bagging and Random Forest improved accuracy over their standard formulations. Their rankings are now reversed: ISLE-RF is often more accurate than ISLE-Bagging, so the extra perturbation was worth it. Finally, the ISLE-based versions can be 20-100 times faster than the standard versions because every tree is built on 5 percent of the data only, and at every iteration, a tree is built with only six terminal nodes, as opposed to fully grown trees.

Although we don't know how much each of the changes made in the algorithms contributed to the improvement in accuracy, it should be clear that they work together: by increasing the "width" of $\hat{r}(p)$, a more "diverse" collection of base learners is obtained, and then the post-process step filters out those base learners that aren't very helpful. But the right width of $\hat{r}(p)$, controlled by the amount of perturbation done, is problem-dependent and unknown in advance. Thus, experimentation is required.

4.5 ADABOOST

Next, the AdaBoost algorithm of Freund and Schapire (1997) is considered. AdaBoost, short for Adaptive Boosting, was initially proposed for 2-class classification problems, and derives its name from its ability to "boost" the performance of "weak" (moderately accurate) base classifiers T_m. Table 4.5 expresses AdaBoost as an ISLE algorithm.

AdaBoost uses an exponential loss function: $L(y, \hat{y}) = \exp(-y \cdot \hat{y})$. $\upsilon = 1$ means AdaBoost implements a sequential sampling approximation to "group" importance where the relevance of base learner T_m is judged in the context of the (fixed) previously chosen base learners $T_1, ..., T_{m-1}$. $\eta = N$, so the data distribution is not perturbed via random sub-sampling; instead, observation weights are used.

The coefficients $\{c_m\}_1^M$ are not estimated with post-processing, but they are not set to $1/M$ either, as was done in the case of Bagging and RandomForest. They are estimated sequentially in the following way: at each iteration, the best T_m is found and then immediately the corresponding c_m is estimated.

In its original formulation, the algorithm was restricted to 2-class problems and the model output was $\hat{y} = \text{sign}\left(F_M(\mathbf{x})\right) = \text{sign}\left(\sum_{m=1}^{M} c_m T_m(\mathbf{x})\right)$. Section 4.5.1 explains why this is a reasonable classification rule under the exponential loss. The algorithm was later extended to regression problems ("Real" AdaBoost (Friedman, J., 2001)). In R, package *gbm* (gbm) implements the AdaBoost exponential loss for 0-1 outcomes.

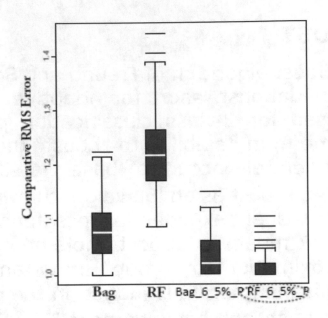

Figure 4.8: Error comparison between classic Bagging and Random Forest with their ISLE-based improved versions (based on (Friedman and Popescu, 2003)). The improved versions use 6-terminal node trees and "post-processing" for fitting the tree coefficients.

Table 4.5: AdaBoost as an ISLE-based algorithm.

- $L(y, \hat{y}) = \exp(-y \cdot \hat{y})$
  $y \in \{-1, 1\}$

- $\upsilon = 1$

- $\eta = N$

- $T_m(\mathbf{x})$: any "weak" learner

- $c_o = 0, \{c_m\}_1^M$: sequential partial regression coefficients

$F_0(\mathbf{x}) = 0$

For $m = 1$ to M {

$\quad (c_m, \mathbf{p}_m) = \arg\min_{c, \mathbf{p}} \sum_{i \in S_m(\eta)} L\big(y_i, F_{m-1}(\mathbf{x}_i) + c \cdot T(\mathbf{x}_i; \mathbf{p})\big)$

$\quad T_m(\mathbf{x}) = T(\mathbf{x}; \mathbf{p}_m)$

$\quad F_m(\mathbf{x}) = F_{m-1}(\mathbf{x}) + \upsilon \cdot c_m \cdot T_m(\mathbf{x})$

}

write $\{c_m, T_m(\mathbf{x})\}_1^M$

Table 4.6 shows the AdaBoost algorithm as presented in its original reference (Freund and Schapire, 1997). Comparing it with the ISLE-based formulation of Table 4.5, the two algorithms don't look alike. However, it can be shown that they are equivalent.

Table 4.6: The AdaBoost algorithm in its original (non-ISLE) formulation.

observation weights : $w_i^{(0)} = 1/N$

For $m = 1$ to M {

    a. Fit a classifier $T_m(\mathbf{x})$ to training data with $w_i^{(m)}$

    b. Compute

$$err_m = \frac{\sum_{i=1}^{N} w_i^{(m)} I(y_i \neq T_m(\mathbf{x}_i))}{\sum_{i=1}^{N} w_i^{(m)}}$$

    c. Compute $\alpha_m = \log((1 - err_m)/err_m)$

    d. Set $w_i^{(m+1)} = w_i^{(m)} \cdot \exp\left[\alpha_m \cdot I(y_i \neq T_m(\mathbf{x}_i))\right]$

}

Output $sign\left(\sum_{m=1}^{M} \alpha_m T_m(\mathbf{x})\right)$

The proof, given in Appendix A, requires showing:

1. Line $p_m = \underset{P}{\arg\min}$ $(\cdot)$ in the ISLE-based algorithm is equivalent to line a. in the AdaBoost algorithm

2. Line $c_m = \underset{c}{\arg\min}$ $(\cdot)$ in the ISLE-based algorithm is equivalent to line c. in the AdaBoost algorithm

3. How the weights $w_i^{(m)}$ are derived/updated.

4.5.1 EXAMPLE

Implementing the AdaBoost algorithm in R is straightforward, as shown in Table 4.7. Figure 4.9 illustrates the evolution of observation weights for the 2-input, 2 class problem introduced at the beginning of this chapter.

Table 4.7: The AdaBoost algorithm in R for 2-class classification.

```
fitAdaBoostTrees <- function(data, M) {
  N <- nrow(data)
  w <- rep(1/N, N)
  alpha <- rep(NA, M)
  treeList <- vector(mode = "list", length = M)
  tree.params <- list(minsplit = 4, minbucket = 4, maxdepth = 2)
  for (m in 1:M) {
    treeList[[m]] <- fitClassTree(data, tree.params, w)
    yHat <- predict(treeList[[m]], data, type = "class")
    i.mistakes <- which(yHat != data$y)
    err <- sum(w[i.mistakes])/sum(w)
    alpha[m]=log((1-err)/err)
    ind <- rep(0, N)
    ind[i.mistakes] <- 1
    if ( m == 1 ) {
      W <- w
    } else {
      W <- cbind(W, w)
    }
    w <- w * exp(alpha[m] * ind)
  }
  return(list(trees=treeList, wghts=W))
}
```

Initially, all the weights are equal and set to $1/N$. As the algorithm evolves, it up-weights the cases that remain in error-especially if they're rare amongst those that are misclassified-and it down-weights the cases that are being recognized correctly. After changing the case weights, and saving the previous model, the algorithm builds a new model using these new case weights. So as a case is misclassified over and over, its relative weight grows, and the cases that the model gets right start to fade in importance. Eventually as you iterate, the points that are in dispute have all the weight and are all along the boundary.

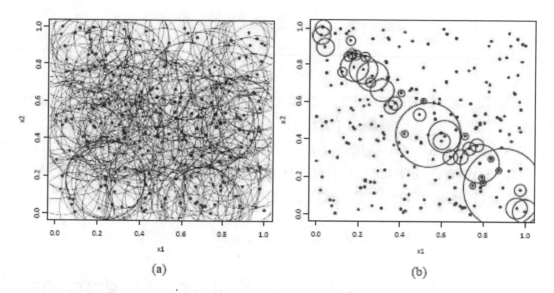

(a)   (b)

Figure 4.9: Evolution of AdaBoost observation weights in a 2-input, 2 class problem that has a linear decision boundary. In (a), all observations have the same weight. In (b), after a few iterations, only points along the decision boundary have a non-zero weight.

4.5.2 WHY THE EXPONENTIAL LOSS?

It is not evident from Table 4.6 that AdaBoost is using an exponential loss. This fact was established in (Friedman, J., 2001) after the algorithm had been proposed and found by practitioners to work well. One advantage of using the exponential loss is implementation convenience. That is, the exponential loss leads to the straightforward algorithm of Table 4.7. With other loss functions that one might like to use, solving the optimization problem involved in finding each T_m (i.e., solving Equation (4.4)) is a more elaborate process.

Putting implementation convenience aside, is there a connection between the function built by AdaBoost and class probabilities? Happily, it can be shown that the algorithm converges towards the "half-log odds" (see proof in Section 4.5.2):

$$F^{\text{AdaBoost}}(\mathbf{x}) = \arg\min_{F(\mathbf{x})} E_{Y|\mathbf{x}}\left(e^{-Y\cdot F(\mathbf{x})}\right)$$
$$= \tfrac{1}{2}\log\frac{\Pr(Y=1|\mathbf{x})}{\Pr(Y=-1|\mathbf{x})}$$

This is a reassuring result; it says that the exponential loss leads to a meaningful population minimizer, which justifies the use of *sign(F)* as the classification rule. It can also be shown that this is the same population minimizer obtained using the (negative) binomial log-likelihood,

$$E_{Y|\mathbf{x}}\left(-l(Y, F(\mathbf{x}))\right) = E_{Y|\mathbf{x}}\left(\log(1 + e^{-Y\cdot F(\mathbf{x})})\right)$$

which is a more familiar loss in other domains. The exponential loss and the (negative) binomial log-likelihood, or binomial deviance, can be seen as continuous approximations to the misclassification loss. This is illustrated in Figure 4.10, which shows them both as functions of the "margin" $y \cdot F$. As indicated in Table 4.5, we are using a –1 *vs* 1 encoding of the response in a 2-class classification problem. If the response y has the same sign as the ensemble output $F(\mathbf{x})$, then the misclassification cost is 0; otherwise, it is 1 (see the blue line).

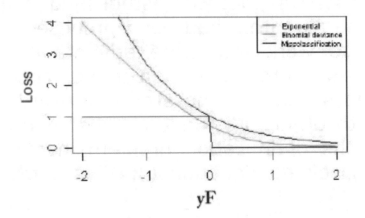

Figure 4.10: The exponential loss and binomial deviance approximations to misclassification loss.

Misclassification loss, however, is discontinuous at $y \cdot F = 0$ and this precludes the use of gradient-descent algorithms to find the minimum risk T_m (Line 3 of Algorithm 4.1). The exponential loss and the binomial deviance are thus "surrogate" criteria used to solve the minimization problem. Both penalize negative margin values (i.e., mistakes) more heavily than they reward increasingly positive ones. The difference is in the degree: binomial deviance increases linearly, whereas exponential loss increases exponentially, thus concentrating more influence on observations with large negative margins. Thus, in noisy situations where there may be mislabeled data, exponential loss will lead to performance degradation.

Finally, the exponential loss has no underlying probability and has no natural generalization to K classes, whereas the binomial deviance does (Friedman, J., 1999).

4.5.3 ADABOOST'S POPULATION MINIMIZER

In this section, we show that $F^{\text{AdaBoost}}(x) = \arg\min_{F(x)} E_{Y|x}\left(e^{-Y \cdot F(x)}\right) = \frac{1}{2}\log\frac{\Pr(Y=1|x)}{\Pr(Y=-1|x)}$. We start by expanding $E_{Y|x}\left(e^{-Y \cdot F(x)}\right)$:

$$E(e^{-yF(x)}|x) = \Pr(Y=1|x) \cdot e^{-F(x)} + \Pr(Y=-1|x) \cdot e^{F(x)}$$

Next, we find the minimum of this function by setting its first derivative to zero and solving for $F(\mathbf{x})$:

$$\frac{\partial E(e^{-yF(x)}|x)}{\partial F(x)} = -\Pr(Y=1|x) \cdot e^{-F(x)} + \Pr(Y=-1|x) \cdot e^{F(x)}$$

thus

$$\frac{\partial E(e^{-yF(x)}|x)}{\partial F(x)} = 0 \Rightarrow \Pr(Y=-1|x) \cdot e^{F(x)} = \Pr(Y=1|x) \cdot e^{-F(x)}$$
$$\Rightarrow \ln \Pr(Y=-1|x) + F(x) = \ln \Pr(Y=1|x) - F(x)$$
$$\Rightarrow 2F(x) = \ln \Pr(Y=1|x) - \ln \Pr(Y=-1|x)$$
$$\Rightarrow F(x) = \frac{1}{2}\ln\frac{\Pr(Y=1|x)}{\Pr(Y=-1|x)}.$$

4.6 GRADIENT BOOSTING

The next seminal work in the theory of Boosting came with the generalization of AdaBoost to any differential loss function. Friedman's Gradient Boosting algorithm (Friedman, J., 2001), with the default values for the algorithm parameters, is summarized in Table 4.8.

Table 4.8: Gradient Boosting as an ISLE-based algorithm.

- General $L(y, \hat{y})$ and y

- $\upsilon = 0.1$

- $\eta = N/2$

- $T_m(x)$: any "weak" learner

- $c_o = \arg\min_c \sum_{i=1}^{N} L(y_i, c)$

- $\{c_m\}_1^M$: "shrunk" sequential partial regression coefficients

$F_0(\mathbf{x}) = c_0$

For $m = 1$ to M {

$\quad (c_m, \mathbf{p}_m) = \arg\min_{c,\mathbf{p}} \sum_{i \in S_m(\eta)} L(y_i, F_{m-1}(\mathbf{x}_i) + c \cdot T(\mathbf{x}_i; \mathbf{p}))$

$\quad T_m(\mathbf{x}) = T(\mathbf{x}; \mathbf{p}_m)$

$\quad F_m(\mathbf{x}) = F_{m-1}(\mathbf{x}) + \upsilon \cdot c_m \cdot T_m(\mathbf{x})$

}

write $\{(\upsilon \cdot c_m), T_m(\mathbf{x})\}_1^M$

The first element of the ensemble is a constant function which depends on the particular loss being used (see Appendix B for an example). $\upsilon > 0$, means Gradient Boosting also implements the sequential sampling approximation to "group" importance; however, compared with AdaBoost's $\upsilon = 1$, more perturbation of the loss is taking place here. As in AdaBoost's case, the coefficients are fitted incrementally, one at every step.

The similarity with the Forward Stagewise Linear Regression procedure of Section 3.3.4, with $\{T_m(\mathbf{x})\}_1^M$ as predictors, should be evident. And the fact that $\upsilon < 1$ means that Gradient Boosting is doing shrinkage implicitly: the coefficients are "shrunk" with a Lasso-type penalty with the shrinkage controlled by υ. This fact is today understood to be one of the key reasons for the superior performance of the algorithm.

The process of instantiating the above template to any differential loss function is covered in Appendix B.

In R, package *gbm* (gbm) implements Gradient Boosting for regression and classification problems using a variety of loss functions (e.g., least squares, absolute-deviations, logistic, etc.).

4.7 MART

MART, short for Multiple Additive Regression Trees, is the name that J. Friedman gave to his implementation of Gradient Boosting when the base learners are trees, which leads to computationally efficient update rules in the gradient descent algorithm required to solve Equation (4.4) (as shown in Appendix B).

When using trees as base learners, one has the option of controlling the size of each tree, which is important for the following reason. The theory of "ANOVA decomposition of a function" tells us that most multivariate functions can be expressed as a constant plus the sum of functions of one variable (main effects), plus the sum of functions of two variables (two-variable interaction effects), and so on. Most functions can then be expressed in the following form:

$$
\begin{aligned}
F(\mathbf{x}) = \ & \sum_j f_j(x_j) \\
& + \sum_{j,k} f_{jk}(x_j, x_k) \\
& + \sum_{j,k,l} f_{jkl}(x_j, x_k, x_l) \\
& + \ldots
\end{aligned}
$$

and usually there is a dominant "interaction order" among its variables. Thus, if the functional model we are using to build our approximation doesn't support the right interaction order, we will be incurring extra bias and/or variance.

If the base learners $T_m(\mathbf{x})$ are J–terminal-node trees, then the interaction order allowed by the approximation can be controlled by adjusting J. For instance, using $J = 2$ means the ensemble can model "main-effects-only" functions because a 2-terminal-node tree is a function of only one variable. Similarly, using $J = 3$ means the ensemble can model target functions of up to two-variable interactions. Again, this is so because a 3-terminal-node tree is a function of at most 2 variables and so on.

Obviously, the optimal value of J should reflect the (unknown) dominant interaction level of the target function. Thus, one must fine-tune the value of J by trying different values and choosing the one that produces the lowest error. As a rule of thumb, however, in most practical problems low-order interactions tend to dominate; i.e., $2 \leq J \leq 8$.

4.8 PARALLEL VS. SEQUENTIAL ENSEMBLES

Figure 4.11 is taken from a Friedman and Popescu (2003) comparative study of over 100 different target functions. The first four algorithms are the same ones discussed in Figure 4.8: Bagging, Random Forest, and their corresponding ISLE –based improvements. The notation "Seq_η_υ%_P" indicates 6-terminal-node trees, "sequential" approximation to importance sampling, η "memory" factor, υ%-size samples without replacement, and post-processing. Thus, MART is equivalent to Seq_0.1_50%.

As in the cases of Bagging and Random Forest, adding the post-processing step of fitting the coefficients $\{c_m\}_0^M$ to the data using regularized regression after the $\{T_m\}_1^M$ have been fitted results in an improvement over MART. The best algorithm also differs from MART in terms of η and υ: a smaller υ means that each T_m is fitted to a smaller data set (increased diversity), which requires smaller values of η (slower learning).

Overall, the "sequential" ISLE algorithms tend to perform better than parallel versions. This is consistent with results observed in classical Monte Carlo integration (Friedman and Popescu, 2003).

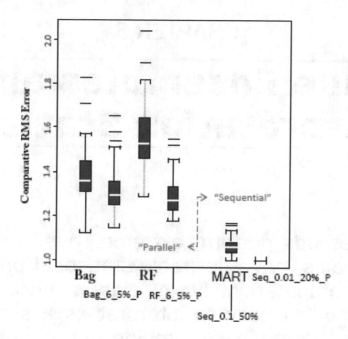

Figure 4.11: Comparison between "Parallel" and "Sequential" ensemble models (adapted from (Friedman and Popescu, 2003)).

## CHAPTER 5

Rule Ensembles and Interpretation Statistics

Ensemble methods perform extremely well in a variety of problem domains, have desirable statistical properties, and scale well computationally. A classic ensemble model, however, is not as easy to interpret as a simple decision tree. In this chapter, we provide an overview of Rule Ensembles (Friedman and Popescu, 2005; Friedman and Bogdan, 2008), a new ISLE-based model built by combining simple, readable rules. While maintaining (and often improving) the accuracy of the classic tree ensemble, the rule-based model is much more interpretable. In this chapter, we will also illustrate recently proposed interpretation statistics which are applicable to Rule Ensembles as well as to most other ensemble types.

Recapping from Chapter 4, an ensemble model of the type we have been considering is a linear expansion of the form

$$F(\mathbf{x}) = c_0 + \sum_{m=1}^{M} c_m f_m(\mathbf{x}) \tag{5.1}$$

where the $\{f_m(\mathbf{x})\}_1^M$ are derived predictors ("basis" functions or "base learners") which capture non-linearities and interactions. Fitting this model to data is a 2-step process:

1. Generate basis functions $\{f_m(x) = f(x; p_m)\}_1^M$, and

2. post-fit to the data via regularized regression to determine the coefficients $\{c_m\}_0^M$.

The resulting model is almost always significantly more accurate than a single decision tree. The simple interpretation offered by a single tree, however, is no longer available. A significant improvement toward interpretability is achieved with "rule" ensembles.

5.1 RULE ENSEMBLES

As discussed in Chapter 2, the functional form of a J-terminal node decision tree can be expressed as a linear expansion of indicator functions:

$$T(\mathbf{x}) = \sum_{j=1}^{J} \hat{c}_j I_{\hat{R}_j}(\mathbf{x})$$

where

$I_A(\mathbf{x})$ is 1 if $\mathbf{x} \in A$ and 0 otherwise,

the $\hat{c}_j$'s are the constants associated with each terminal node,

and the $\hat{R}_j$'s are the regions in $\mathbf{x}$-space defined by the terminal nodes.

In the case of terminal nodes that are at a depth greater than 1, the corresponding $I_{\hat{R}_j}(\mathbf{x})$ functions can themselves be expressed as a product of indicator functions. For instance, $\hat{R}_1$ in Figure 2.3 is defined by the product of $I(x_1 > 22)$ and $I(x_2 > 27)$. Thus, $I_{\hat{R}_1}(\mathbf{x})$ is equivalently expressed by the following conjunctive rule:

"if $x_1 > 22$ and $x_2 > 27$ then 1 else 0"

Table 5.1 lists the rules for the other regions induced by the tree of Figure 2.3. In the ensemble model of Equation

(5.1), the base learners $f_m(\mathbf{x})$ typically have been trees, but one could use the rules $r_j(\mathbf{x})$ instead. Because each rule is a simple readable statement about attributes of $\mathbf{x}$, then a model using these rules can be more interpretable.

Table 5.1: The regions in input ($\mathbf{x}$) space generated by a decision tree expressed as a set of binary rules (adapted from (Hastie et al., 2001)).

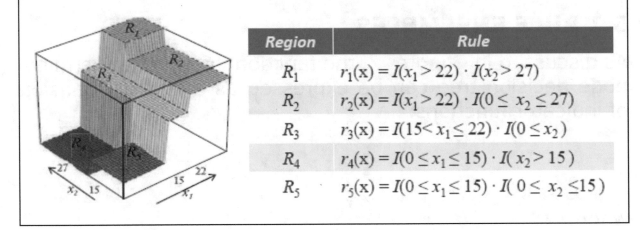

Region	Rule
R_1	$r_1(\mathbf{x}) = I(x_1 > 22) \cdot I(x_2 > 27)$
R_2	$r_2(\mathbf{x}) = I(x_1 > 22) \cdot I(0 \leq x_2 \leq 27)$
R_3	$r_3(\mathbf{x}) = I(15 < x_1 \leq 22) \cdot I(0 \leq x_2)$
R_4	$r_4(\mathbf{x}) = I(0 \leq x_1 \leq 15) \cdot I(x_2 > 15)$
R_5	$r_5(\mathbf{x}) = I(0 \leq x_1 \leq 15) \cdot I(0 \leq x_2 \leq 15)$

A rule ensemble is still a piecewise constant model and linear targets are still be problematic as they were with single trees. But it is possible to combine base learners from different families. Thus, the non-linear rules can be complemented with purely linear terms. The resulting model has the form:

$$F(\mathbf{x}) = a_0 + \sum_m a_m r_m(\mathbf{x}) + \sum_j b_j x_j \qquad (5.2)$$

In terms of the ISLE framework (Algorithm 4.1), we need some approximate optimization approach to solve:

$$\mathbf{p}_m = \arg\min_{\mathbf{P}} \sum_{i \in S_m(\eta)} L\left(y_i, \ F_{m-1}(\mathbf{x}_i) + r(\mathbf{x}_i; \mathbf{p})\right)$$

where $\mathbf{p}_m$ are the split definitions for rule $r_m(\mathbf{x})$. But instead of solving this optimization problem directly, it is possible to take advantage of a decision tree ensemble in the following way: once the set $\{T_m(x)\}_1^M$ has been built, each tree $T_m(\mathbf{x})$ can be decomposed into its rules $r_j(\mathbf{x})$, and let those be the base learners that are being combined together during the coefficient estimation step.

In the case of shallow trees, as commonly used with Boosting, regions corresponding to nonterminal nodes can also be included. If all nodes are counted, a J-terminal node tree generates $2 \times (J - 1)$ rules. Once the rules have been generated, the next step is to fit the corresponding coefficients using the same linear regularized procedure discussed in Chapters 3 and 4:

$$(\{\hat{a}_k\}, \{\hat{b}_j\}) = \underset{\{a_k\}, \{b_j\}}{\arg \min} \sum_{i=1}^{N} L\left(y_i, \; a_0 + \sum_{k=1}^{K} a_k r_k(x_i) + \sum_{j=1}^{P} b_j x_{ij}\right)$$
$$+ \lambda \left(\sum_{k=1}^{K} |a_k| + \sum_{j=1}^{P} |b_j|\right)$$

where,

- $p \le n$ indicates the number of input predictors that are of continuous type, and which are desired to be included as purely linear terms. It is often sensible to replace these x_j variables by their "winzorized" (Winsorize) versions.

- M is the number of trees in the ensemble.

- $K = \sum_{m=1}^{M} 2 \times (J_m - 1)$ denotes the total number of rules.

As in the case of tree ensembles, tree size, J, controls rule "complexity:" a J-terminal node tree can generate rules involving up to $(J - 1)$ variables. Thus, modeling J-order interactions requires rules with J or more components.

5.2 INTERPRETATION

Chapter 2 defined the Predictive Learning problem and identified two types of modeling tasks: "predictive" and "descriptive." A predictive classification model has the goal of assigning a set of observations into one of two or more classes. For example, a predictive model can be used to score a credit card transaction as fraudulent or legitimate. A descriptive model, on the other hand, has the additional goal of describing the classification instead of just determining it. In semiconductor manufacturing, for example, descriptive models are used to identify and understand defect causes. In these descriptive tasks, the ability to derive interpretations from the resulting classification model is paramount.

Over the last few years, new summary statistics have been developed to interpret the models built by ensemble methods. These fall into three groups:

1. *Importance scores*: to answer the question of which particular variables are the most relevant. Importance scores quantify the relative influence or contribution of each variable in predicting the response. In the manufacturing domain, for example, datasets can contain thousands of predictor variables, and importance scores allow the analyst to focus the defect diagnosis effort on a smaller subset of them.

2. *Interaction statistic*: to answer the question of which variables are involved in interactions with other variables and to measure the strength and degrees of those interactions. In the manufacturing domain case again, interaction statistics could show that a combination of a particular machine being used for a certain step is a cause of defects but only when operated at a certain time of the day.

3. *Partial dependence plots*: to understand the nature of the dependence of the response on influential inputs.

For example, does the response increase monotonically with the values of some predictor x_j? Partial dependence plots allow visualizing the model as a function of, say, two of the most important variables at a time (while averaging over all other variables).

Most of the math related to the calculation of these interpretation statistics can be applied to any base learner, but they are easier to compute for trees (as shown in Sections 5.2.2 through 5.2.4 below).

5.2.1 SIMULATED DATA EXAMPLE

To illustrate the interpretation methodology, consider the following simple artificial example (borrowed from (Breiman et al., 1993)). There are ten predictor variables generated according to the following rule:

$$p(x_1 = -1) = p(x_1 = 1)\ 1/2$$
$$p(x_2 = -1) = p(x_2 = 1)\ 1/2$$

$$p(x_m = -1) = p(x_m = 0) = p(x_m = 1) = 1/3 \quad m = 3,...,10 \quad (5.3)$$

and the response variable according to:

$$y = \begin{cases} 3 + 3 \cdot x_2 + 2 \cdot x_3 + x_4 + z & x_1 = 1 \\ -3 + 3 \cdot x_5 + 2 \cdot x_6 + x_7 + z & x_1 = -1 \end{cases} \quad (5.4)$$

Thus, x_1 and x_2 take two values, -1 and 1, with equal probability, and the other eight variables take three values, -1, 0, and 1, also with equal probability. The response variable depends on x_1 through x_7, but not on x_8, x_9, or x_{10}. Therefore, x_8, x_9, and x_{10} are irrelevant inputs that will be present in the input vectors **x**. The response variable definition also indicates that there is an interaction between x_1 and x_2, x_1 and x_3, x_1 and x_4, but there is no interaction among them. Similarly, there is an interaction between x_1

and x_5, x_1 and x_6, x_1 and x_7, with no interaction among x_5,x_6, and x_7. Finally, z represents noise added to the response variable and has a normal distribution $z \sim N(0, 2)$.

The goal is to see whether or not one can recover the target rule definition using the ensemble interpretation methodology.

We generated a small training data set with $N = 200$. Figure 5.1 shows a single tree fitted to this data. The tree correctly splits on x_1 first, suggesting the greater importance of this variable. Then, it splits on x_2 and x_5 at the second level, suggesting that these variables come next in terms of relevance in predicting the response. The tree splits on x_3 and x_7 after that. Irrelevant variables x_8, x_9, and x_{10} are not present in any tree split.

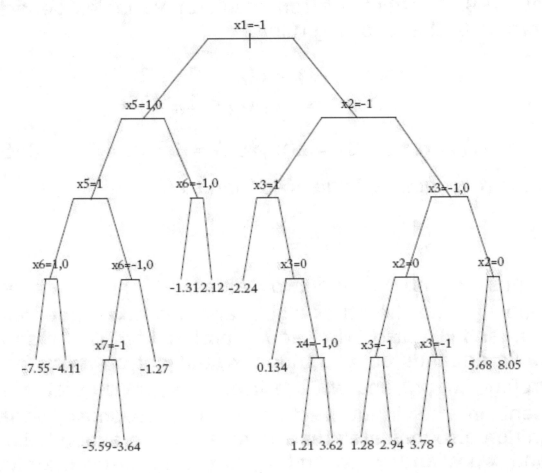

Figure 5.1: Decision tree for data generated according to Equations (5.3) and (5.4) in the text.

Table 5.2 shows the eight globally most important rules in a rule ensemble model fit to the same data. For every rule, its *importance, coefficient,* and *support* is listed. Since the rules are binary functions, the *coefficient* values, a_m in Equation (5.2), represent the change in predicted value if the rule is satisfied (or "fires"). *Support* refers to the fraction of the data to which the rule applies. Rule *importance* is based on the coefficient magnitude and support of the rule (defined in Section 5.2.2). Thus, rule 1 applies to ~30% of the data and is interpreted as cases for which $x_1 = 1$ and x_2 is either 0 or 1, tend to have higher response (given the magnitude and sign of its coefficient). The first rule suggests x_1 is interacting with x_2, the second rule suggests x_1 is interacting with x_5, and so on.

Table 5.2: Rule ensemble for the data generated according to Equations (5.3) and (5.4) in the text.

Importance	Coefficient	Support	Rule
100	1.80	0.30	$x_1 = 1$ and $x_2 \in \{0,1\}$
95	-2.25	0.14	$x_1 = -1$ and $x_5 \in \{-1\}$
83	-1.59	0.24	$x_1 = 1$ and $x_2 \in \{-1,0\}$ and $x_3 \in \{-1,0\}$
71	-1.00	0.35	$x_1 = -1$ and $x_6 \in \{-1\}$
62	1.37	0.17	$x_1 = 1$ and $x_2 \in \{0,1\}$ and $x_3 \in \{0,1\}$
57	1.25	0.35	$x_1 = -1$ and $x_6 \in \{-1,0\}$
46	1.00	0.11	$x_1 = -1$ and $x_5 \in \{1\}$ and $x_7 \in \{0,1\}$
42	0.91	0.51	$x_1 = 1$ and $x_4 \in \{1\}$

Figure 5.2(a) shows the variable importance score computed using the R package *rule-fit* (RuleFit). The

importance score ranks all the input variables according to the strength of their effect on the model. As expected, the most important variable is x_1, and it is followed by x_2 and x_5 with roughly the same relevance. Then, x_3 and x_6, followed by x_4 and x_7. Finally, x_8, x_9, and x_{10}, which don't play a role in the target, appear last. It's interesting to note that, while none of these variables have any impact on the single tree in Figure 5.1, they have non-zero importance, as can be seen in the graph. This means the variables do show up in some rules in the ensemble of trees but the corresponding coefficients are very small.

Figure 5.3(b) shows the interaction for each variable. It indicates which variables are likely to be involved in interactions with others. The red bars correspond to the reference, or "null," distribution values corresponding to the hypothesis of no interaction effects. Thus, the height of each yellow bar reflects the value of the interaction statistic in excess of its expected value.

The reference distributions are computed using a "bootstrap" method. The idea is to generate random data sets that are similar to the original training data, repeatedly compute the interaction statistic on these artificial data, and compare the resulting values with those obtained from the original data (Friedman and Popescu, 2005).

One sees that x_1 has been identified as the variable interacting the strongest (most involved in interactions) then x_2 and x_5, followed by x_3 and x_6 to a lesser extent. Going back to the target function definition, Equation (5.4), and examining the magnitude of the different coefficients there,one sees a good match with this ranking. The participation of x_4 and x_7 in the interactions, with x_1 in the target, is not detected—i.e., it is measured to be too weak. Having identified which variables are involved in interactions, the next question in the model interpretation process is finding those variables with which they are interacting. The two-variable interaction statistic is intended to answer this question.

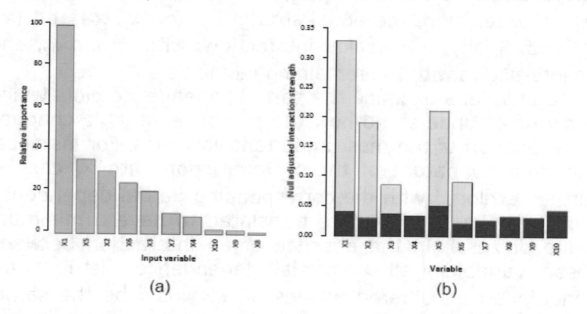

(a)   (b)

Figure 5.2: Variable Importance and Interaction Statistic for the data generated according to Equations (5.3) and (5.4).

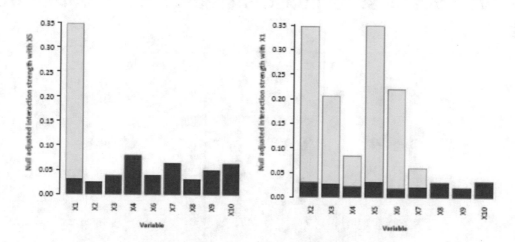

Figure 5.3: Two-variable Interaction Statistic for the data generated according to Equations (5.3) and (5.4).

Figure 5.3 shows the two-variable interaction statistics $(X_5, *)$ and $(X_1, *)$. It is seen that x_5 has been identified to be only interacting with x_1, which corresponds with how we constructed the data. And with which variables is x_1 interacting? We see the (x_1, x_2) interaction as strong as the (x_1, x_5) one, then, the equal strength pairs (x_1, x_3) and (x_1, x_6) and, finally, the weaker interactions with x_4 and x_7, and no interaction with the remaining variables.

Lastly, let's examine the partial dependence plots which allow us to understand how the response variable changes as a function of the most important variables. For instance, the detailed nature of the x_1 interaction with x_5 can be further explored with the corresponding partial dependence plot (see Figure 5.4 – here translated to have a minimum value of zero). In the absence of an interaction between these variables, all x_5 partial dependence distributions, conditioned on different values of x_1, would be the same. Here, one sees that when $x_1 = 1$, x_5 has no effect on the response. The distribution for $x_1 = -1$ is very different and captures the essence of the interaction effect between these two variables: response increases as x_5 varies from – 1 to 1.

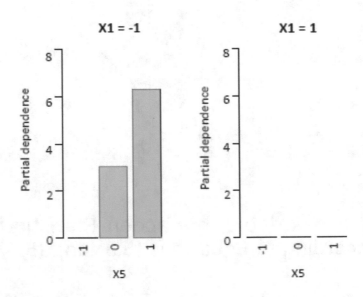

Figure 5.4: Two-variable Partial Dependence example for the simulated data generated according to Equations (5.3) and (5.4).

5.2.2 VARIABLE IMPORTANCE

How are variable importance scores such as those shown in Figure 5.2 computed? The single tree variable importance measure is given by the following expression (Breiman et al., 1993):

$$\Im(x_l; T) = \sum_{t \in T} \hat{I}\left(v(t), s_{v(t)}\right) \cdot I\left(v(t) = l\right) \qquad (5.5)$$

which is simply the sum of the "goodness of split" scores $\hat{I}$ (see Section 2.1) whenever x_l is used in the split definition of an internal node t in the tree T. Since only the relative magnitudes of the $\Im(x_l)$ are interesting, often the following normalized measure is reported instead:

$$\text{Imp}(x_j) = 100 \cdot \Im(x_j) / \max_{1 \le l \le n} \Im(x_l)$$

For a tree ensemble, Equation (5.5) can be easily generalized to an average across all the trees in the ensemble:

$$\Im(x_j) = \frac{1}{M} \sum_{m=1}^{M} \Im(x_j; T_m)$$

In the case of a rule ensemble, which is a linear model of rules and linear "terms," it's possible to define "global" and "local" (term) importance scores:

- Global term importance: as generally used in linear models, defined as the absolute value of the coefficient of the standardized predictor:

    - Rule term: $\Im_k = \left|\hat{a}_k\right| \cdot \sqrt{s_k \cdot (1 - s_k)}$
      where s_k denotes the support of the rule; i.e., $s_k = \frac{1}{N} \sum_{i=1}^{N} r_k(x_i)$.

    - Linear term: $\Im_j = \left|\hat{b}_j\right| \cdot \text{std}(x_j)$.

- Local term importance: defined at *each* point **x** as the absolute change in $\hat{F}(\mathbf{x})$ when the term is removed from the ensemble:

  ○ Rule term: $\Im_k(\mathbf{x}) = |\hat{a}_k| \cdot |r_k(\mathbf{x}) - s_k|$.

  ○ Linear term: $\Im_j(\mathbf{x}) = |\hat{b}_j| \cdot |x_j - \bar{x}_j|$.

Note that a term has an importance score at a particular point **x** even if the rule doesn't fire, or the particular predictor $x_j = 0$, for the given case. The two local term importance scores can be combined to arrive at a local variable importance score:

$$\Im(x_l; \mathbf{x}) = \Im_l(\mathbf{x}) + \sum_{x_l \in r_k} \Im_k(\mathbf{x}) / \text{size}(r_k)$$

which can be averaged over any subset of the input space to obtain a "regional" importance score. For example, the variable ranking could be computed for the region where all **x**'s are positive. It can be shown that, if the variable rankings $\Im(x_i; \mathbf{x})$ are averaged over the entire input space, the global metrics are recovered (Friedman and Popescu, 2004).

5.2.3 PARTIAL DEPENDENCES

The purpose of the partial dependence plots, such as the one shown in Figure 5.4, is to visualize the effect on $\hat{F}(\mathbf{x})$ of a small subset of the input variables $x_s \subset \{x_1, x_2,..., x_n\}$ – e.g., a subset that has been deemed interesting based on the variable importance score.

Since $\hat{F}(\mathbf{x})$ also depends on the other ("complement") variables $\mathbf{x}_c$ – i.e., $\hat{F}(\mathbf{x}) = \hat{F}(\mathbf{x}_s, \mathbf{x}_c)$ and $\mathbf{x}_s \cup \mathbf{x}_c = \{x_1, x_2,..., x_n\}$, such visualization is only possible after accounting for the (average) effects of $\mathbf{x}_c$. Thus, the partial dependence on $\mathbf{x}_s$ is defined by (Hastie et al., 2001):

$$\hat{F}(\mathbf{x}) = \hat{F}(\mathbf{x}_S, \mathbf{x}_C)$$

which is approximated by:

$$\hat{F}_S(\mathbf{x}_S) = \frac{1}{N} \sum_{i=1}^{N} F(\mathbf{x}_S, x_{iC})$$

where $\{x_{1C}, \ldots, x_{NC}\}$ are the values of $\mathbf{x}_C$ occurring in the training data.

5.2.4 INTERACTION STATISTIC

How are interaction scores such as those shown in Figure 5.3 computed? If x_j and x_k do not interact, then $\hat{F}(\mathbf{x})$ can be expressed as the sum of two functions:

$$\hat{F}(\mathbf{x}) = f_{\setminus j}(\mathbf{x}_{\setminus j}) + f_{\setminus k}(\mathbf{x}_{\setminus k})$$

with $f_{\setminus j}(\mathbf{x}_{\setminus j})$ not depending on x_j and $f_{\setminus k}(\mathbf{x}_{\setminus k})$ not depending on x_k. Thus, the partial dependence of $\hat{F}(\mathbf{x})$ on $\mathbf{x}_S = \{x_j, x_k\}$ can also be decomposed as the sum of two functions:

$$\hat{F}_{j,k}(x_j, x_k) = \hat{F}_j(x_j) + \hat{F}_k(x_k) \tag{5.6}$$

namely, the sum of the respective partial dependencies. This immediately suggests a way to test for the presence of an (x_j, x_k) interaction: check whether the equality of Equation (5.6) holds. More formally, the two-variable interaction statistic is defined by (Friedman and Popescu, 2005):

$$H_{jk}^2 = \sum_{i=1}^{N} \left[\hat{F}_{j,k}(x_{ij}, x_{ik}) - \hat{F}_j(x_{ij}) - \hat{F}_k(x_{ik}) \right]^2 \Big/ \sum_{i=1}^{N} \hat{F}_{j,k}^2(x_{ij}, x_{ik})$$

with $H_{jk}^2 \approx 0$ indicating $\hat{F}_{j,k}(x_j, x_k) \approx \hat{F}_j(x_j) + \hat{F}_k(x_k)$ and thus no interaction between x_j and x_k. Likewise, $H_{jk}^2 \gg 0$ would indicate the presence of an interaction.

Similarly, if x_j does not interact with any other variable, then $\hat{F}(\mathbf{x})$ can be expressed as the sum:

$$\hat{F}(\mathbf{x}) = F_j(x_j) + F_{\setminus j}(\mathbf{x}_{\setminus j})$$

with $F_j(x_j)$ being a function only of x_j and $F_{\setminus j}(\mathbf{x}_{\setminus j})$ a function of all variables except x_j. Thus, a single-variable interaction statistic to test whether x_j interacts with *any* other variable can be defined by (Friedman and Popescu, 2005):

$$H_j^2 = \sum_{i=1}^{N} \left[\hat{F}(\mathbf{x}_i) - \hat{F}_j(x_{ij}) - \hat{F}_{\setminus j}(\mathbf{x}_{i \setminus j}) \right]^2 \Bigg/ \sum_{i=1}^{N} \hat{F}^2(\mathbf{x}_i) .$$

5.3 MANUFACTURING DATA EXAMPLE

In this section, we present the application of rule ensembles, and the associated interpretation methodology discussed above, to an analysis of semiconductor manufacturing data (Seni et al.,2007). The data consists of 2045 observations, each described by 680 input variables. Observations correspond to wafers, and the variables encode manufacturing attributes such as the machine name and time of various processing steps. There were missing values present in the data. The response variable corresponds to "yield" (the number of die in a wafer that pass several electrical tests) which ranges from 60% to 96% in this data. The goal is to build a model that can help establish if certain machines (or combinations of them) at a certain time, and/or at certain process steps, cause low yield.

Table 5.3 summarizes the estimated test error results using three models: main effects only (i.e., using single-variable rules in the ensemble), a single tree, and an rule ensemble with interactions allowed. For the single tree, the tree size is given. For the ensemble model, the number of terms (rules and linear) with nonzero coefficients is shown.

Table 5.3: Average absolute prediction error (estimated via cross-validation) for different models applied to a semiconductor data set.

	Error	*Size*
Main effects only model	0.91	
Single tree	0.44	32 terminal nodes
Model with interactions	0.35	220 terms

The average absolute prediction error for the ensemble model is significantly lower than the corresponding error for an additive model restricted to main effects only. Thus, there is good evidence for interaction effects being present. The single tree model had 32 terminal nodes and a depth of 7 – i.e., it is not so easy to read. In this case, the ensemble model improves the single-tree's error by ~20% (.44 to .35).

Table 5.4 shows the three globally most important rules in the ensemble model. Figure 5.5(a) shows the relative importance of the ten most important input variables (out of the 680 given ones), as averaged over all predictions. Variable PE.3880.4350.ILTM444, which is also present in the top rules, figures prominently. An analysis (not shown) of the partial dependence of yield on this variable, reveals that wafers using YEQUIP707, or YEQUIP709, at step 3880.4350 tend to have noticeably lower yield.

Table 5.4: Top three rules based on global importance for the ensemble model for the yield data.

Imp	Coef	Supp	Rule
100	1.58	0.25	PE.1550.1000.LTH203 ∈ {LEQUIP754} & PE.3560.4720.ILT112 ∉ {YEQUIP704, YEQUIP706, YEQUIP710, YEQUIP711} & PE.3880.4350.ILTM444 ∉ {YEQUIP702, YEQUIP706, YEQUIP707, YEQUIP709} & TI.1850.1805.WETETCH13 ≤ 38620
89	-1.37	0.27	PE.3300.0400.WETCFF21 ∈ {SEQUIP702} & PE.3670.4200.CMP552 ∉ {PEQUIP702} & TI.3230.2115.INS711 ≥ 38620
71	-1.09	0.29	PE.2100.1175.ION621 ∉ {IEQUIP703} & PE.2450.1040.WETS23 ∉ {CEQUIP704} & PE.3970.4200.CMP554 ∉ {PEQUIP706, PEQUIP707}

The analysis of interaction effects can now be focused on the smaller set of variables deemed most relevant. The yellow bars in Figure 5.5(b) show the values of the statistic used to test whether a specified variable interacts with any other variable. The red bars correspond to the reference (null) distribution values. Thus, the height of each yellow bar reflects the value of the interaction statistic in excess of its expected value under the null hypothesis of no interaction effects.

Although the strengths of the interaction effects shown in Figure 5.5(b) are not large, at least two of the fifteen most influential variables appear to be involved in interactions with other variables. After identifying those variables that interact with others – e.g., PE.2100.1175.ION621 and PE.2550.1000.LTH233 above, we need to determine the particular other variables with which they are interacting.

The values of the two-variable interaction strength statistic for PE.2550.1000.LTH233 are shown in Figure 5.6. Here, see that PE.2550.1000.LTH233 dominantly interacts with PE.2997.0100.WETC755.

The detailed nature of this interaction can be further explored with the corresponding partial dependence plot (see Figure 5.7). In the absence of an interaction between these variables, all PE.2550.1000.LTH233 partial dependence distributions, conditioned on different values of PE.2997.0100.WETC755, would be the same.

Here we see similar distributions when PE.2997.0100.WETC755 takes the values DEQUIP701 (Figure 5.7(a)) and DEQUIP702 (Figure 5.7(b)) (with one PE.2550.1000.LTH233 value not represented in one case). The distribution for PE.2997.0100.WETC755 = DEQUIP703 (Figure 5.7(c)) is fairly different from the others, suggesting it captures the essence of the interaction effect between these two variables. (Yield is lower throughout in this case.) The impact of this interaction can be further visualized using "wafer map" plots: Figure 5.8(a) shows average yield for the wafers using LEQUIP701 at step 2550.1000 or using DEQUIP703 at step 2997.0100. Figure 5.8(b) shows average yield for the "complement" set of wafers. A ~7.5% yield loss is observed.

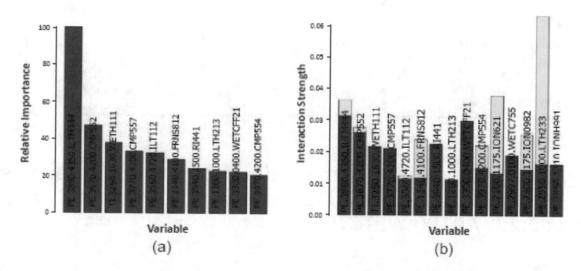

(a)   (b)

Figure 5.5: Interpretation statistics for the yield data. In (a), input variable relative (global) importance. In (b), total interaction strengths in excess of expected null value for the top 15 input variables.

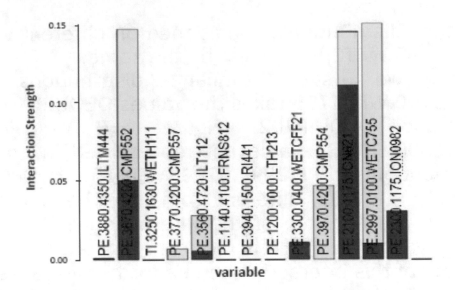

Figure 5.6: Two-variable interaction strengths of variables interacting with PE.2550.1000.LTH233.

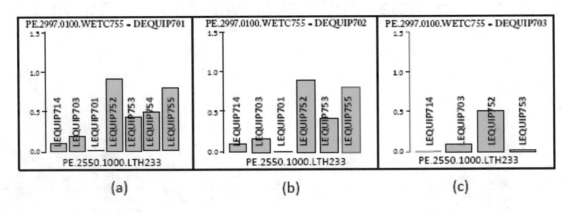

Figure 5.7: Joint partial dependence on variables PE.2550.1000.LTH233 (found earlier to be a relevant input) and PE.2997.0100.WETC755.

5.4 SUMMARY

Ensemble methods in general, and boosted decision trees in particular, constitute one of the most important advances in machine learning in recent years. In the absence of detailed *a priori* knowledge of the problem at hand, they provide superior performance. A number of interpretation tools have been developed that, while applicable to other algorithms, are often easiest to compute for trees. With the introduction of rule ensembles, the interpretability of the ensemble model has been further significantly improved. Together, they provide a powerful way to solve a variety of regression and classification problems.

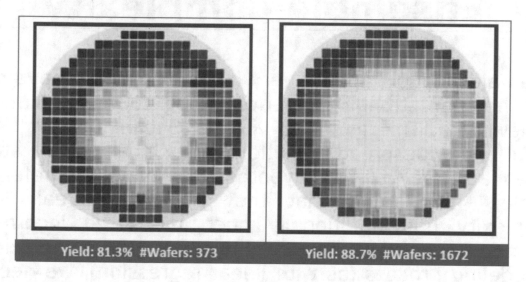

Figure 5.8: Wafer map representation of yield. Minimum values are represented by darker colored die, graduating to lighter die for higher values. In (a), average yield for the 373 wafers using LEQUIP701 at step 2550.1000 or using DEQUIP703 at step 2997.0100. In (b), average yield for the remaining 1672 wafers.

## CHAPTER 6

Ensemble Complexity

Ensemble[1] models appear very complex, yet we have seen how they can strongly outperform their component models on new data. This seems to violate "Occam's Razor" – the useful and widespread analytic doctrine that can be stated "when accuracy of two hypotheses is similar, prefer the simpler." We argue that the problem is really that complexity has traditionally been measured incorrectly. Instead of counting parameters to assess the complexity of a modeling process (as with linear regression), we need to instead measure its *flexibility* – as by Generalized Degrees of Freedom, GDF (Ye, J., 1998). By measuring complexity according to a model's behavior rather than its appearance, the utility of Occam's Razor is restored. We'll demonstrate this on a two-dimensional decision tree example where the whole (an ensemble of trees) has less GDF complexity than any of its parts.

6.1 COMPLEXITY

One criticism of ensembles is that interpretation is much harder than with a single model. For example, decision trees have properties so attractive that, second to linear regression (LR), they are the modeling method most widely employed, despite having the worst accuracy of the major algorithms. Bundling trees into an ensemble makes them competitive on this crucial property, though at a serious loss in interpretability. (The Rule Ensembles and companion interpretation methodology of Chapter 5 make ensembles much more interpretable, but some loyal tree users will still prefer the simplicity of a single tree.) Note that an ensemble of trees can itself be represented as a tree, as it produces a piecewise constant response surface[2]. But the tree equivalent to an ensemble can have vastly more nodes than the component trees; for example, a bag of M "stumps" (single-split binary trees) requires up to 2^M leaves to be represented by a single tree (when all the splits are on different variables).

Indeed, Bumping (Tibshirani and Knight, 1999a) was designed to get some of the benefit of bagging without requiring multiple models, in order to retain some interpretability. It builds competing models from bootstrapped datasets and keeps only the one with least error on the original data. This typically outperforms, on new data, a model built simply on the original data, likely due to a bumped model being robust enough to do well on two related, but different datasets. But, the expected improvement is greater with ensembles.

Another criticism of ensembles – more serious to those for whom an incremental increase in accuracy is worth a multiplied decrease in interpretability – is the concern that their increased complexity will lead to overfit, that is, inaccuracy on new data. In fact, not observing ensemble overfit in practical applications has helped throw into doubt, for many, the Occam's Razor axiom that generalization is hurt by complexity. (This and other critiques of the axiom are argued in an award-winning paper by Domingos, P. (1998).)

But, are ensembles truly complex? They appear so; but do they *act* so? The key question is how we should measure complexity. For LR, one can merely count coefficients (to measure the degrees of freedom of the data used up by the process of parameter estimation), yet this is known to fail for nonlinear models. It is possible for a single parameter in a nonlinear model to have the influence of less than a single linear parameter, or greater than several – e.g., it is estimated that each parameter in Multivariate Adaptive Regression Splines uses three effective degrees of freedom (Friedman, J., 1991; Owen, A., 1991). The under-linear case can occur with say, a neural network that hasn't trained long enough to pull all its weights into play. The over-linear case is more widely known. For example, Friedman and Silverman (1989) note the following: "The results of Hastie and Tibshirani (1985), together with those of Hinkley, D. (1969, 1970) and Feder, P. (1975), indicate that the number of degrees of freedom associated with nonlinear least squares regression can be considerably more than the number of parameters involved in the fit."

The number of parameters and their degree of optimization is not all that contributes to a model's complexity or its potential for overfit. The model form alone doesn't reveal the *extent of the search for* structure. For example, the winning model for the 2001 Knowledge Discovery and Data Mining (KDD) Cup settled on a linear model employing only three variables. This appears simple, by definition. Yet, the data had 140,000 candidate variables, constrained by only 2,000 cases. Given a large enough ratio of unique candidate variables to cases, even a blind search will very likely find some variables that look explanatory when there is no true relationship. As Hjorth, U. (1989) warned, "...the evaluation of a selected model can not be based on that model alone, but requires information about the class of models and the selection procedure." We thus need to employ model selection metrics that include the effect of model selection! (Note that the valuable tool of cross-validation (CV) can be used effectively, but only if *all* steps in the modeling process are automated and included inside the CV loop.)

There is a growing realization that complexity should be measured not just for a model, but for an entire modeling *procedure*, and that it is closely related to that procedure's *flexibility*. For example, the Covariance Inflation Criterion (Tibshirani and Knight, 1999b) fits a model and saves the output estimates, $\hat{y}_i$, then randomly shuffles the output variable, y, re-runs the modeling procedure, and measures the covariance between the new and old estimates. The greater the change (adaptation to randomness, or flexibility) the greater the complexity penalty needed to restrain the model from overfit (see Section 3.3). Somewhat more simply, Generalized Degrees of Freedom, GDF (Ye, J., 1998) adds random noise to the output variable, re-runs the modeling procedure, and measures the changes to the estimates. Again, the more a modeling procedure adapts to match the added noise the more flexible (and therefore more complex) its model is deemed to be.

The key step in both – a randomized loop around a modeling procedure – is reminiscent of the Regression Analysis Tool (Farway, J., 1991), which measured, through resampling, the robustness of results from multi-step automated modeling. Whereas at that time sufficient resamples of a two-second procedure took two days, increases in computing power have made such empirical measures much more practical.

6.2 GENERALIZED DEGREES OF FREEDOM

For LR, the degrees of freedom, K, equal the number of coefficients, though this does not extrapolate well to nonlinear regression. But, there exists another definition that does:

$$GDF(F, D) = \sum_{i=1}^{N} \Delta \hat{y}_i / \Delta y_i$$

where[3]

$$\Delta y_i = y_{e_i} - y_i, \text{ and } \Delta \hat{y}_i = \hat{y}_{e_i} - \hat{y}_i$$
$$y_{e_i} = y_i + \varepsilon_i \text{ with } \varepsilon_i \sim N(0, \sigma^2)$$
$$\hat{y}_i = F_y(\mathbf{x}_i) \text{ is the output of model } F \text{ trained on data } D = \{y_i, \mathbf{x}_i\}_1^N$$
$$\hat{y}_{e_i} = F_{y_e}(\mathbf{x}_i) \text{ is the output of model } F \text{ trained on data } D_e = \{y_{e_i}, \mathbf{x}_i\}_1^N \qquad (6.1)$$

GDF is thus defined to be the sum of the average sensitivity of each fitted value, $\hat{y}_i$, to perturbations in its corresponding target, y_i. As it may be calculated experimentally for all algorithms, GDF provides a valuable way to measure and compare the flexibility (and parameter power) of any modeling methods.

In practice, Ye suggests generating a table of perturbation sensitivities, then employing a "horizontal" method of calculating GDF, as diagrammed in Figure 6.1. In that table, the rows correspond to the observations (cases), the columns correspond to the randomly-perturbed replicant data sets D_e, and each cell holds the perturbed output, y_e, and its estimate, $\hat{y}_e$, for one case and sample. A case's sensitivity (of its estimate to its perturbing noise), m_i, is estimated by fitting a LR to $\Delta \hat{y}_i$ vs. Δy_i using the row of data for case i. (Since y_i and $\hat{y}_i$ are constant, the LR simplifies to be $\hat{y}_{ei}$ vs. y_{ei}.) The GDF is then the sum of these slopes, m_i. This "horizontal" estimate seems to be more robust than that obtained by the "vertical" method of averaging the value obtained for each column of data (i.e., the GDF estimate for each perturbation dataset).

6.3 EXAMPLES: DECISION TREE SURFACE WITH NOISE

We take as a starting point for our tests the two-dimensional piecewise constant surface used to introduce GDF (Ye, J., 1998), shown in Figure 6.2.

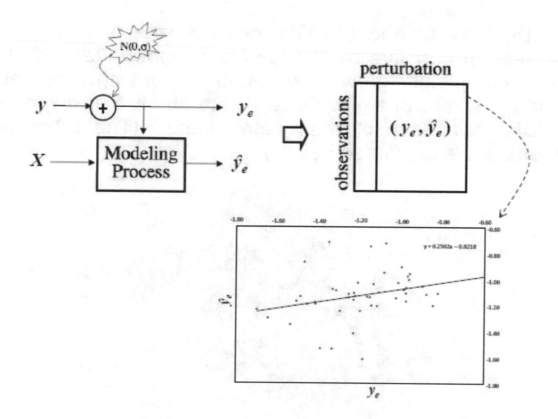

Figure 6.1: Diagram of GDF computation process.

It is generated by (and so it can be perfectly fit by) a decision tree with five terminal (leaf) nodes (i.e., four splits), whose smallest structural change (level change) is 0.5. Figure 6.3 illustrates the surface after Gaussian noise N (0, 0.5) has been added, and Figure 6.4 shows a sample made of 100 random observations of that space. This tree+noise data is the dataset, $D_e = \{y_{e_i}, x_i\}_1^{100}$, employed for the experiments. For GDF perturbations, we employed 50 such random samples (replicates), $D_e^1, \ldots, D_e^{50}$, where each added to y Gaussian noise, $N(0, 0.25)$, having half the standard deviation of the noise already in the training data[4].

Figure 6.5 shows the GDF vs. K (number of parameters) sequence for LR models, single trees, and ensembles of five trees (and two more sequences described below). Confirming theory, the GDF for the LR models closely matches the number of terms, K. For single decision trees (*Tree 2d*) of different sizes, K (maximum number of splits), the GDF grew at about 3.67 times the rate of K. Bagging (See Section 4.3) five trees together (*Bagged Tree 2d*), the rate of complexity growth was 3.05. Surprisingly perhaps, the bagged trees of a given size, K, are about a fifth simpler, by GDF, than each of their components! (The other two sequences are explained below.)

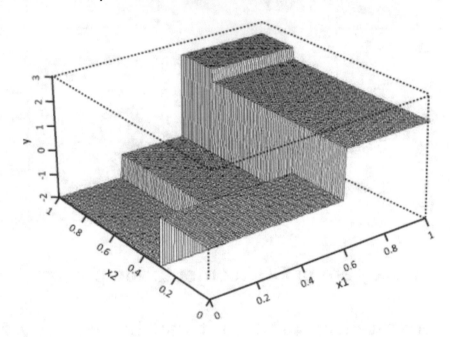

Figure 6.2: (Noiseless version of) two-dimensional tree surface used in experiments (after Ye, J. (1998)).

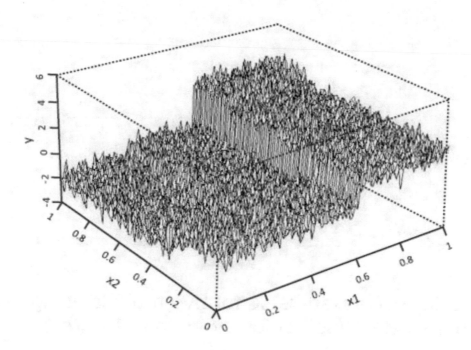

Figure 6.3: Tree surface of Figure 6.2 after adding $N(0, 0.5)$ noise.

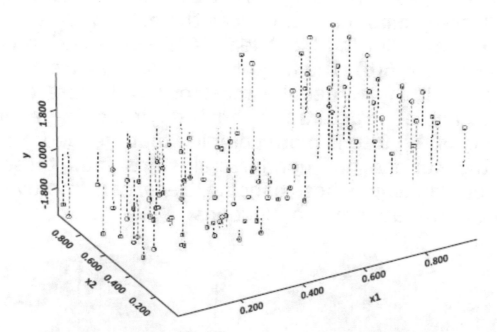

Figure 6.4: Sample of 100 observations from Figure 6.3 (dotted lines connect points to zero plane).

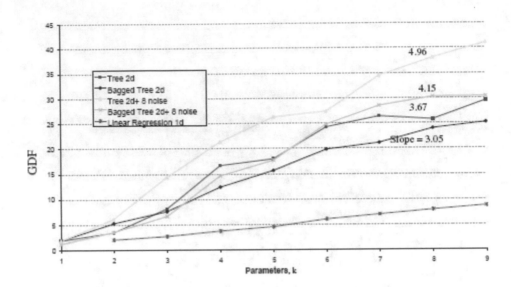

Figure 6.5: GDF vs. splits, *K*, for five models using from one to nine parameters (splits) for the data of Figure 6.4.

Figure 6.6 illustrates two of the surfaces in the sequence of bagged trees. Bagging five trees limited to four leaf nodes (three splits) each produces the estimation surface of Figure 6.6(a). Allowing eight leaves (seven splits) produces that of Figure 6.6(b). The bag of more complex trees creates a surface with finer detail – most of which here does not relate to actual structure in the underlying data-generating function, as the tree is more complex than needed. For both bags, the surface has gentler stair-steps than those of a lone tree, revealing how bagging trees can improve their ability to estimate smooth functions.

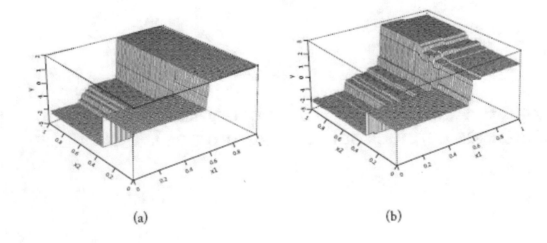

(a)   (b)

Figure 6.6: Surface of bag of five trees using (a) three splits, and (b) seven splits.

Expanding the experiment (after Ye, J. (1998), we appended eight random candidate input variables, $x_3,..., x_{10}$, to **x**, to introduce *selection noise*, and re-ran the sequence of individual and bagged trees. Figures 6.7(a) and 6.7(b) illustrate two of the resulting bagged surfaces (projected onto the space of the two relevant inputs), again for component trees with three and seven splits, respectively. The structure in the data is clear enough for the under-complex model to avoid using the random inputs, but the over-complex model picks some up. Figure 6.5 shows the GDF progression for the individual and bagged trees with ten candidate inputs (*Tree 2d+ 8 noise*, and *Bagged tree 2d+ 8 noise*, respectively). Note that the complexity slope for the bagged tree ensemble (4.15) is again less than that for its components (4.96). Note also that the complexity for each ten-input experiment is greater than its corresponding two-input one. Thus, even though one cannot tell – by looking at a final model using only the relevant inputs x_1 and x_2 – that random variables were considered, their presence increases the chance for overfit, and this is appropriately reflected in the GDF measure of complexity.

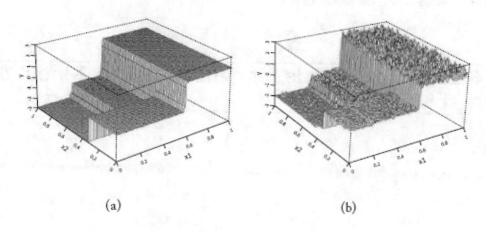

(a)   (b)

Figure 6.7: Surface of bag of five trees fit to a ten-dimensional input data with 8 irrelevant (noise) inputs. In (a), three split trees were used. In (b), seven split trees were used. Surface is drawn on the plane of the two relevant inputs.

6.4 R CODE FOR GDF AND EXAMPLE

Table 6.1 displays code for generating the two-dimensional data of Figure 6.3.

To generate the data we simply do:

```
data2d.100 <- gen2dData(N=100, noise.seed=321, noise.sd=0.5)
```

and to compute GDF for a single decision tree:

```
gdf <- GDF(data2d.100, modelTrainer=treeModelTrainer, replicates=50,
           noise.seed=321, noise.sd=0.25)
```

where a decision tree is built by:

```
treeModelTrainer <- function(data, K) {
  library(rpart)
  tree <- rpart(y ~ . , data=data,control=
                      rpart.control(minsplit=2, cp=0.0001))
  ## Prune tree back to desired number of splits
  i.alpha <- which.min(abs(tree$cptable[,"nsplit"] - K))
  alpha.K <- tree$cptable[i.alpha,"CP"]
  tree.p <- prune(tree, cp = alpha.K)
  return(tree.p)
}
```

Table 6.1: Sample R code for generating the 2-input data of Figure 6.3.

```
gen2dData <- function(N, noise.seed, noise.sd) {
  set.seed(noise.seed)
  x1 <- runif(N, 0, 1)
  x2 <- runif(N, 0, 1)
  y <- rep(NA, N)
  for (i in 1:N) {
    if ( x1[i] > .6 ) {
      if ( x2[i] > .8 ) {
        y[i] <- rnorm(1, 2.5, noise.sd)
      } else {
        y[i] <- rnorm(1, 2, noise.sd)
      }
    } else {
      if ( x2[i] < .3 ) {
        y[i] <- rnorm(1, 0, noise.sd)
      } else {
        if ( x1[i] > .3 ) {
          y[i] <- rnorm(1, -1, noise.sd)
        } else {
          y[i] <- rnorm(1, -2, noise.sd)
        }
      }
    }
  }
  return(data.frame(x1, x2, y))
}
```

6.5 SUMMARY AND DISCUSSION

Bundling competing models into ensembles almost always improves generalization – and using different algorithms as the perturbation operator is an effective way to obtain the requisite diversity of components. Ensembles appear to increase complexity, as they have many more parameters than their components; so, their ability to generalize better seems to violate the preference for simplicity summarized by Occam's Razor. Yet, if we employ GDF – an empirical measure of the *flexibility* of a modeling *process* – to measure complexity, we find that ensembles can be simpler than their components. We argue that when complexity is thereby more properly measured, Occam's Razor is restored.

Under GDF, the more a modeling process can match an arbitrary change made to its output, the more complex it is. The measure agrees with linear theory, but can also fairly compare very different, multi-stage modeling processes. In our tree experiments, GDF increased in the presence of distracting input variables, and with parameter power (i.e., decision trees use up more degrees of freedom per parameter than does LR), GDF is expected to also increase with search thoroughness, and to decrease with use of Bayesian parameter priors, or parameter shrinkage, or when the structure in the data is clear relative to the noise. Additional observations (constraints) may affect GDF either way.

Table 6.2: Sample R code to compute Generalized Degrees of Freedom (GDF) for a given data set and model. Argument modelTrainer is a function which can be invoked to train a model of some desired type; K stands for number of terms, number of splits, etc., in the model. Argument n.rep is the number of times to replicate (perturb) the data. The last two arguments control the noise added in the perturbations.

```
GDF <- function(data, modelTrainer, K, n.rep, noise.seed, noise.sd) {
  N <- nrow(data)
  ## Create the noise to be added over the data replicates
  set.seed(noise.seed)
  perturbations <- matrix(rnorm(N * n.rep, 0, noise.sd), nrow=N)
  ye.mat <- matrix(data$y, nrow=N, ncol=n.rep, byrow=F) + perturbations

  ## Train a model on input data; store yHat
  base_model <- modelTrainer(data, K)
  yHat <- predict(base_model, data)

  yeHat.mat <- matrix(NA, nrow=N, ncol=n.rep)

  data_perturbed <- data
  for (i in 1:n.rep) {
    data_perturbed$y <- ye.mat[,i]
    ## Train a model on perturbed data; evaluate on input x
    model_perturbed <- modelTrainer(data_perturbed, K)
    yeHat.mat[,i] <- predict(model_perturbed, data)
  }

  GDF.m <- c(NA, N)
  for (i in 1:N) {
    lmodel <- lm(yeHat.mat[i,] - yHat[i] ~ perturbations[i,])
    GDF.m[i] <- lmodel$coefficients[2]
  }
  GDF <- sum(GDF.m)
  return(GDF)
}
```

Lastly, case-wise (horizontal) computation of GDF has an interesting by-product: a measure of the complexity contribution of each case. Figures 6.8(a) and 6.8(b) illustrates these contributions for two of the single-tree models of Figure 6.5 (having three and seven splits, respectively). The under-fit tree results of Figure 6.8(a) reveal only a few observations to be complex, that is, to lead to changes in the model's estimates when perturbed by random noise. (Contrastingly, the complexity is more diffuse for the results of the overfit tree in Figure 6.8(b).) A future modeling algorithm could recursively seek such *complexity contribution outliers* and focus its attention on the local model structure necessary to reduce them, without increasing model detail in regions which are stable.

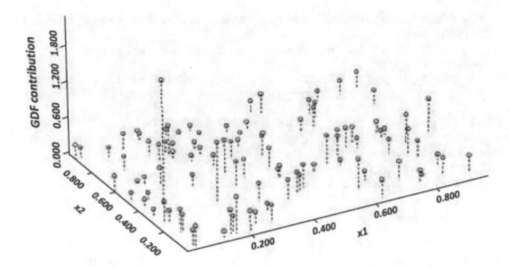

Figure 6.8: (a): Complexity contribution of each sample for bag of five trees using three splits.

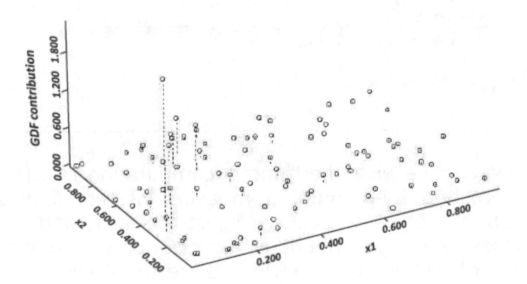

Figure 6.8: (b): Complexity contribution of each sample for bag of five trees using seven splits.

[1]This chapter is based on Elder, J. (2003). The Generalization Paradox of Ensembles, *Journal of Computational and Graphical Statistics* 12, No. 4: 853-864.

[2]The decision boundary induced by a weighted sum of trees, as from bagging or boosting, is piecewise constant and so can be represented by a single tree. To build it, generate training data from a fine grid of input points and run it through the ensemble to generate the target variable from the model, then fit a tree perfectly to the resulting data.

[3]We enjoyed naming the perturbed output, ($y_e = y + \varepsilon$) after GDF's inventor, Ye.

[4]Where the degree of noise in a dataset can be estimated, it is a rule of thumb for the perturbation magnitude to be half as large.

AdaBoost Equivalence to FSF Procedure

In this appendix we show that the AdaBoost algorithm presented in Table 4.6 is equivalent to the Forward Stagewise Fitting (FSF) procedure of Table 4.5. At every iteration of the algorithm, we need to solve:

$$(c_m, \mathbf{p}_m) = \arg\min_{c, \mathbf{p}} \sum_{i=1}^{N} L\left(y_i,\ F_{m-1}(\mathbf{x}_i) + c \cdot T(\mathbf{x}_i; \mathbf{p})\right)$$

and since $L(y, \hat{y}) = \exp(-y.\hat{y})$, we can write

$$(c_m, \mathbf{p}_m) = \arg\min_{c, \mathbf{p}} \sum_{i=1}^{N} \exp\left(-y_i \cdot F_{m-1}(\mathbf{x}_i) - c \cdot y_i \cdot T(\mathbf{x}_i; \mathbf{p})\right)$$

$$= \arg\min_{c, \mathbf{p}} \sum_{i=1}^{N} w_i^{(m)} \cdot \exp\left(-c \cdot y_i \cdot T(\mathbf{x}_i; \mathbf{p})\right) \qquad (A.1)$$

where $w_i^{(m)} = e^{-y_i F_{m-1}(\mathbf{x}_i)}$. Since $w_i^{(m)}$ doesn't depend on c or $\mathbf{p}$, it can be regarded as an observation weight. A solution to Equation (A.1) can be obtained in two steps: - Step 1: given c, solve for $T_m = T(\mathbf{x}; \mathbf{p}_m)$:

$$T_m = \arg\min_{T} \sum_{i=1}^{N} w_i^{(m)} \cdot \exp\left(-c \cdot y_i \cdot T(\mathbf{x}_i)\right) \qquad (A.2)$$

- Step 2: given T_m, solve for c

$$c_m = \arg\min_{c} \sum_{i=1}^{N} w_i^{(m)} \cdot \exp\left(-c \cdot y_i \cdot T_m(\mathbf{x}_i)\right) \qquad (A.3)$$

We need to show that solving Step 1 above is equivalent to line a. in the AdaBoost algorithm, and that the solution to Step 2 is equivalent to line c. in the same algorithm (Table

4.5). That is, we need to show that T_m is the classifier that minimizes the weighted error rate. We start by expanding Equation (A.2) above:

$$T_m = \arg\min_{T} \sum_{i=1}^{N} w_i^{(m)} \cdot \exp(-c \cdot y_i \cdot T(x_i)) = \arg\min_{T}\left[e^{-c} \cdot \sum_{y_i=T(x_i)} w_i^{(m)} + e^{c} \cdot \sum_{y_i \neq T(x_i)} w_i^{(m)}\right]$$

which follows from the following simple facts:

- $y \in \{-1, 1\}, T(x) \in \{-1, 1\}$
- $y = T(x) \Rightarrow \exp(-c \cdot y \cdot T(x)) = e^{-c}$
- $y \neq T(x) \Rightarrow \exp(-c \cdot y \cdot T(x)) = e^{c}$

and, thus the derivation can be continued as:

$$\arg\min_{T}\left[e^{-c} \cdot \sum_{y_i=T(x_i)} w_i^{(m)} + e^{c} \cdot \sum_{y_i \neq T(x_i)} w_i^{(m)}\right]$$

$$= \arg\min_{T}\left[e^{-c} \cdot \sum_{i=1}^{N} w_i^{(m)} - e^{-c} \cdot \sum_{y_i\neq T(x_i)} w_i^{(m)} + e^{c} \cdot \sum_{y_i \neq T(x_i)} w_i^{(m)}\right]$$

$$= \arg\min_{T}\left[e^{-c} \cdot \sum_{i=1}^{N} w_i^{(m)} + (e^{c} - e^{-c}) \cdot \sum_{y_i \neq T(x_i)} w_i^{(m)}\right]$$

$$= \arg\min_{T}\left[(e^{c} - e^{-c}) \cdot \sum_{i=1}^{N} w_i^{(m)} I(y_i \neq T(x_i)) + e^{-c} \cdot \sum_{i=1}^{N} w_i^{(m)}\right]$$

$(e_c - e^{-c})$ is constant and $e^{-c} \cdot \sum_{i=1}^{N} w_i^{(m)}$ doesn't depend on T, thus the last line above leads to: $= \arg\min_{T}\left[\sum_{i=1}^{N} w_i^{(m)} I(y_i \neq T(x_i))\right]$

In other words, the T that solves Equation (A.2) is the classifier that minimizes the weighted error rate.

Next, we need to show that the constant c_m that solves Equation (A.3) is $c_m = \frac{1}{2}\log\frac{1 - err_m}{err_m}$ as required by line c. in the AdaBoost algorithm. We start by expanding Equation (A.3):

$$c_m = \arg\min_{c}\left[(e^{c} - e^{-c}) \cdot \sum_{i=1}^{N} w_i^{(m)} I(y_i \neq T_m(x_i)) + e^{-c} \cdot \sum_{i=1}^{N} w_i^{(m)}\right]$$

and computing and setting to zero its derivative with respect to c:

$$\frac{\partial}{\partial c}\left((e^{c} - e^{-c}) \cdot \sum_{i=1}^{N} w_i^{(m)} I(y_i \neq T_m(x_i)) + e^{-c} \cdot \sum_{i=1}^{N} w_i^{(m)}\right)$$

$$= (e^{c} + e^{-c}) \cdot \sum_{i=1}^{N} w_i^{(m)} I(y_i \neq T_m(x_i)) - e^{-c} \cdot \sum_{i=1}^{N} w_i^{(m)} = 0$$

which follows simply from the derivative of the Exponential function. Thus,

$$e^c \cdot \sum_{i=1}^{N} w_i^{(m)} I(y_i \neq T_m(x_i)) + e^{-c} \cdot \sum_{i=1}^{N} w_i^{(m)} I(y_i \neq T_m(x_i)) - e^{-c} \cdot \sum_{i=1}^{N} w_i^{(m)} = 0$$

dividing by e^{-c}:

$$e^{2c} \cdot \sum_{i=1}^{N} w_i^{(m)} I(y_i \neq T_m(x_i)) + \sum_{i=1}^{N} w_i^{(m)} I(y_i \neq T_m(x_i)) - \sum_{i=1}^{N} w_i^{(m)} = 0$$

$$\Rightarrow e^{2c} \cdot \sum_{i=1}^{N} w_i^{(m)} I(y_i \neq T_m(x_i)) = \sum_{i=1}^{N} w_i^{(m)} - \sum_{i=1}^{N} w_i^{(m)} I(y_i \neq T_m(x_i))$$

$$\Rightarrow e^{2c} = \frac{\sum_{i=1}^{N} w_i^{(m)} - \sum_{i=1}^{N} w_i^{(m)} I(y_i \neq T_m(x_i))}{\sum_{i=1}^{N} w_i^{(m)} I(y_i \neq T_m(x_i))}$$

$$\Rightarrow c = \frac{1}{2} \ln \frac{\sum_{i=1}^{N} w_i^{(m)} - \sum_{i=1}^{N} w_i^{(m)} I(y_i \neq T_m(x_i))}{\sum_{i=1}^{N} w_i^{(m)} I(y_i \neq T_m(x_i))}$$

Separately, we have that the AdaBoost algorithm calls for $c_m = \frac{1}{2} \ln \frac{1 - \text{err}_m}{\text{err}_m}$ with $\text{err}_m = \frac{\sum_{i=1}^{N} w_i^{(m)} I(y_i \neq T_m(x_i))}{\sum_{i=1}^{N} w_i^{(m)}}$. Thus,

$$c_m = \frac{1}{2} \ln \frac{1 - \frac{\sum_{i=1}^{N} w_i^{(m)} I(y_i \neq T_m(x_i))}{\sum_{i=1}^{N} w_i^{(m)}}}{\frac{\sum_{i=1}^{N} w_i^{(m)} I(y_i \neq T_m(x_i))}{\sum_{i=1}^{N} w_i^{(m)}}}$$

$$= \frac{1}{2} \ln \frac{\frac{\sum_{i=1}^{N} w_i^{(m)} - \sum_{i=1}^{N} w_i^{(m)} I(y_i \neq T_m(x_i))}{\sum_{i=1}^{N} w_i^{(m)}}}{\frac{\sum_{i=1}^{N} w_i^{(m)} I(y_i \neq T_m(x_i))}{\sum_{i=1}^{N} w_i^{(m)}}}$$

$$= \frac{1}{2} \ln \frac{\sum_{i=1}^{N} w_i^{(m)} - \sum_{i=1}^{N} w_i^{(m)} I(y_i \neq T_m(x_i))}{\sum_{i=1}^{N} w_i^{(m)} I(y_i \neq T_m(x_i))}$$

which is the same as Equation (A.4) above, so the equivalence between the two algorithms is established.

Gradient Boosting and Robust Loss Functions

In this appendix we illustrate the process of instatiating the Gradient Boosting algorithm of Table 4.8 to a particular differentiable loss function.

We need to solve:

$$(c_m, \mathbf{p}_m) = \arg\min_{c,\mathbf{p}} \sum_{i=1}^{N} L\left(y_i, F_{m-1}(\mathbf{x}_i) + c \cdot T(\mathbf{x}_i; \mathbf{p})\right) \tag{B.1}$$

This is done in two steps:

- Step 1: given c, solve for $T_m = T(\mathbf{x}; \mathbf{p}_m)$:

$$\mathbf{p}_m = \arg\min_{\mathbf{p}} \sum_{i=1}^{N} L\left(y_i, F_{m-1}(\mathbf{x}_i) + c \cdot T(\mathbf{x}_i; \mathbf{p})\right) \tag{B.2}$$

- Step 2: given T_m, solve for c:

$$c_m = \arg\min_{c} \sum_{i=1}^{N} L\left(y_i, F_{m-1}(\mathbf{x}_i) + c \cdot T(\mathbf{x}_i; \mathbf{p}_m)\right) \tag{B.3}$$

Solving Equation (B.2) for robust loss functions $L(y,\hat{y})$ such as the absolute loss, Huber loss, binomial deviance, etc., requires use of a "surrogate," more convenient, criterion which is derived from analogy to numerical optimization in function space.

The minimization problem of Equation (B.2) can be simply stated as "find the function f that has minimum risk" – i.e., $\hat{f} = \arg\min_f R(f)$. Each possible f can be viewed as a "point" in $\Re^N$ – i.e., $\mathbf{f} = \langle f(\mathbf{x}_1), f(\mathbf{x}_2),\dots, f(\mathbf{x}_N)\rangle$, and gradient-descent (Duda et al., 2001) can be used in this space to locate the minimum. This is illustrated in Figure B.1 with ρ_m being the step size and $\nabla_m R$ being the gradient vector:

$$\nabla_m R = \begin{bmatrix} \partial R/\partial f(\mathbf{x}_1) \\ \dots \\ \partial R/\partial f(\mathbf{x}_N) \end{bmatrix}_{f=f_{m-1}} = \begin{bmatrix} \partial L\,(y_1, f(\mathbf{x}_1))\,/\partial f(\mathbf{x}_1) \\ \dots \\ \partial L\,(y_N, f(\mathbf{x}_N))\,/\partial f(\mathbf{x}_N) \end{bmatrix}_{f=f_{m-1}}$$

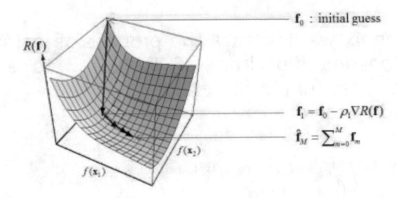

$\mathbf{f}_0$: initial guess

$\mathbf{f}_1 = \mathbf{f}_0 - \rho_1 \nabla R(\mathbf{f})$

$\hat{\mathbf{f}}_M = \sum_{m=0}^{M} \mathbf{f}_m$

Figure B.1: Gradient-descent illustration. The "risk" criterion $R(\mathbf{f})$, is plotted as a function of $\mathbf{f}$ evaluated at two training data points $\mathbf{x}_1$ and $\mathbf{x}_2$. Starting with an initial guess, $\mathbf{f}(\mathbf{x}_0)$, a sequence converging towards the minimum of $R(\mathbf{f})$ is generated by moving in the direction of "steepest" descent —i.e., along the negative of the gradient.

One difficulty, however, is that these $\mathbf{f}$'s are defined on the training data $\mathbf{x}_1$, $\mathbf{x}_2,\dots$, $\mathbf{x}_N$ only. To obtain a function that is defined for all $\mathbf{x}$'s, we can choose the base learner $T(\mathbf{x}; \mathbf{p})$ that is most "parallel" to $\nabla R(\mathbf{f})$ (this is illustrated in Figure B.2) with θ being the angle between the two vectors $T(\mathbf{x}; \mathbf{p})$ and $-\nabla R(\mathbf{f})$. Taking advantage of the geometric interpretation of correlation, we write:

$$\cos\theta = \mathrm{corr}\left(\{-\nabla R\,(f(\mathbf{x}_i))\}_{i=1}^N, \ \{T(\mathbf{x}_i; \mathbf{p})\}_{i=1}^N\right)$$

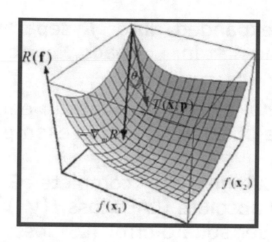

Figure B.2: Surrogate-loss illustration. The base-learner most parallel to the negative gradient vector is chosen at every step of the gradient-descent algorithm.

and the most highly correlated $T(\mathbf{x}; \mathbf{p})$ is given by the solution to: $\mathbf{p}_m = \arg\min_{\beta,\mathbf{p}} \sum_{i=1}^{N} \left(-\nabla_m R\left(f(\mathbf{x}_i)\right) - \beta \cdot T(\mathbf{x}_i; \mathbf{p}) \right)^2$.

Thus, given a differentiable loss function $L(y, \hat{y})$, solving Equation (B.2) simply entails:

1. Set "pseudo response" $\tilde{y}_i = -\left. \frac{\partial L(y_i, \hat{y}_i)}{\partial \hat{y}_i} \right|_{\hat{y}=\hat{y}_{m-1}}$.
   For instance, if $L(y, \hat{y}) = |y - \hat{y}|$, the absolute loss, then $\tilde{y}_i = \text{sign}(y_i - \hat{y}_i)$.

2. Solve the least-squares minimization: $\mathbf{p}_m = \arg\min_{\beta,\mathbf{p}} \sum_{i=1}^{N} \left(\tilde{y}_i - T(\mathbf{x}_i; \mathbf{p}) \right)^2$. If the T's are decision trees, this is just fitting a regression tree to the data $\{\tilde{y}_i, \mathbf{x}_i\}$, using squared-error loss, something we know how to do (see Chapter 2).

Solving Equation (B.3) can be done as outlined in Appendix A, namely by computing and setting to zero the derivative of risk with respect to c, $\partial R(c)/\partial c$. But this step also simplifies itself in the case where the T_m's are trees. Since adding a J–terminal-node tree $T_m(\mathbf{x})$ to the ensemble model is like adding J separate (basis) functions and because the terminal node regions are disjoint, Equation

(B.3) can be expanded into J separate minimization problems, one in each terminal node:

$$\hat{\gamma}_{jm} = \underset{x_i \in \hat{R}_{jm}}{median} \{y_i - F_{m-1}(x_i)\}_1^N \qquad 1 \leq j \leq J.$$

that is, the optimal constant update in each terminal node region. Using the absolute loss, for example, this is simply

$$\hat{\gamma}_{jm} = \underset{x_i \in \hat{R}_{jm}}{median} \{y_i - F_{m-1}(x_i)\}_1^N \qquad 1 \leq j \leq J.$$

Figure B.3 outlines the complete Gradient Boosting algorithm for the absolute (LAD) loss $L(y,\hat{y}) = |y - \hat{y}|$. LAD is often preferred over square-error (LS) loss, $L(y, \hat{y}) = (y - \hat{y})^2$, because it offers resistance to outliers in y. Since trees are already providing resistance to outliers in **x**, an LAD-based ensemble is attractive for regression problems.

Line 1 initializes the ensemble to the best constant function. Line 2 computes the pseudo-response $\tilde{y}_i$, and Line 3 fits a least-squares tree to this response. In Line 4, the constants associated with regions $\hat{R}_{jm}$ of the fitted tree in Line 3 get overwritten. Finally, in Line 5, a "shrunk" version of the new base-learner gets added to the ensemble.

$$F_0(\mathbf{x}) = median\{y_i\}_1^N$$

For $m = 1$ to M {

    // Step 1: find $T_m(\mathbf{x})$

$$\tilde{y}_i = sign(y_i - F_{m-1}(\mathbf{x}_i))$$

$$\{\hat{R}_{jm}\}_1^J = J - \text{terminal node LS-regression tree}\left(\{\tilde{y}_i, \mathbf{x}_i\}_1^N\right)$$

    // Step 2: find coefficients

$$\hat{\gamma}_{jm} = \underset{x_i \in \hat{R}_{jm}}{median}\{y_i - F_{m-1}(\mathbf{x}_i)\}_1^N \qquad j = 1 \ldots J$$

    // Update expansion

$$F_m(\mathbf{x}) = F_{m-1}(\mathbf{x}) + \upsilon \cdot \sum_{j=1}^J \hat{\gamma}_{jm} I\left(\mathbf{x}_i \in \hat{R}_{jm}\right)$$

}

Figure B.3: The Gradient Boosting algorithm for LAD loss.

Bibliography

Barron, R. L., Mucciardi, A. N., Cook, A. N., Craig, J. N., and Barron, A. R. (1984). Adaptive learning networks: Development and application in the United States of algorithms related to GMDH. In *Self-Organizing Methods in Modeling: GMDH Type Algorithms*, S. J. Farlow (ed.), pp. 25–65. New York: Marcel Dekker. 1.1

Bishop, C.M. (1995). *Neural Networks for Pattern Recognition*. Oxford University Press. 2, 4

Breiman, L, Friedman, J.H., Olshen, R., and Stone, C. (1993). *Classification and Regression Trees*. Chappman & Hall/CRC. 1.2, 2, 2.1, 2.2, 3.3.1, 3.3.1, 5.2.1, 5.2.2

Breiman, L. (1995). "Better Subset Regression Using the Nonnegative Garrote." *Technometrics* 37, no. 4, 373–384. DOI: 10.2307/1269730

Breiman, L. (1996). "Bagging predictors." *Machine Learning* **26**(2), 123–140. DOI: 10.1007/BF00058655 1, 1.1, 4.3

Breiman, L. (1998). "Arcing Classifiers." *Annals of Statistics* 26, no. 2, 801–849.

Breiman, L. (2001). *Random Forests, random features*. Berkeley: University of California. 1.1, 4.4

Coifman, R.R., Meyer, Y., and Wickerhauser, V. (1992). "Wavelet analysis and signal processing." In *Wavelets and their Applications*, 153–178. Jones & Bartlett Publishers. 4

data_mining_methods. March (2007). http://www.kdnuggets.com/polls/2007/data_mining_methods.htm 2.2

Domingos, P. (1998). Occam's Two Razors: The Sharp and the Blunt, *Proceedings of the 5th International Conference on Knowledge Discovery and Data Mining*, AAAI Press, New York. 6.1

Duda, Hart, and Stork. *Pattern Classification*, 2nd ed., Wiley, 2001. B

Efron, B., Hastie, T., Johnstone, I. and Tibshirani, R. (2004). "Least Angle Regression." *Annals of Statistics* 32, no. 2, 407–499. 1.2, 3.3.4

Elder, J. F. (1993). *Efficient Optimization through Response Surface Modeling: A GROPE Algorithm*. Dissertation, School of Engineering and Applied Science, University of Virginia, Charlottesville. 1.3

Elder, J. F. (1996a). A review of *Machine Learning, Neural and Statistical Classification* (eds. Michie, Spiegelhalter & Taylor, 1994), *Journal of the American Statistical Association* **91**(433), 436–437. 1

Elder, J. F. (1996b). Heuristic search for model structure: The benefits of restraining greed. In D. Fisher & M. Lenz (Eds.), *Learning from Data: Artificial Intelligence and Statistics*, Chapter 13, New York: Springer-Verlag. 1, 1

Elder, J. F., and Lee, S. S. (1997). *Bundling Heterogeneous Classifiers with Advisor Perceptrons*. University of Idaho Technical Report, October, 14. 1, 1.1, 1, 1.2

Elder, J. F. and Ridgeway, G. (1999). "Combining Estimators to Improve Performance: Bundling, Bagging, Boosting, and Bayesian Model Averaging." *Tutorial at 5th International Conference on Knowledge Discovery & Data Mining*. San Diego, CA, Aug. 15. 4.3

Elder, J. F., and Brown, D. E. (2000). Induction and polynomial networks. In *Network Models for Control and Processing*. M. D. Fraser (ed.), pp. 143–198. Portland: Intellect. 1.1

Elder, J. (2003). "The Generalization Paradox of Ensembles." *Journal of Computational and Graphical Statistics* 12, no. 4, 853–864. DOI: 10.1198/1061860032733 1

faraway. CRAN - Package. http://cran.r-project.org/web/packages/faraway 3.3.5

Faraway, J.J. (1991). On the Cost of Data Analysis,Technical Report, Dept. Statistics, UNC, Chapel Hill. 6.1

Feder, P.I. (1975). The Log Likelihood Ratio in Segmented Regression, *The Annals of Statistics* **3**: 84–97. DOI: 10.1214/aos/1176343000 6.1

Freund, Y. and Schapire, R.E. (1997). "A decision theoretical generalization of on-line learning and an application to boosting." *Journal of Computer and systems Sciences* 55, no. 1, 133–68. DOI: 10.1006/jcss.1997.1504 4.5, 4.5

Freund,Y., and Shapire,R.E.(1996). Experiments with a newboosting algorithm. *Machine Learning: Proceedings of the 13th International Conference*, July, 148–156. 1, 1.1

Friedman, J.H. (1999). "Stochastic gradient boosting." Statistics, Stanford University. DOI: 10.1016/S0167-9473(01)00065-2 1.1, 4.5.2

Friedman, J. H. (2008). "Fast sparse regression and classification." Statistics, Stanford University. 1.2, 3.3.4

Friedman, J. and Popescu, B. E. (2004). "Gradient directed regularization for linear regression and classification." Statistics, Stanford University. 1.2, 3.3.4, 4.4, 5.2.2

Friedman, J. and Popescu, B. E. (2003). "Importance Sampled Learning Ensembles." Statistics, Stanford University. 4, 4.2, 4.3, 4.8, 4.8, 4.11

Friedman, J. and Popescu, B. E. (2005). "Predictive learning via rule ensembles." Statistics, Stanford University. 5, 5.2.1, 5.2.4

Friedman, J. (2001). "Greedy function approximation: the gradient boosting machine." *Annals of Statistics* 29, no. 2, 1189–1232. DOI: 10.1214/aos/1013203451 1.1, 4.5, 4.5.2, 4.6

Friedman, J.H. (1991). Multivariate Adaptive Regression Splines, *Annals of Statistics* **19**. DOI: 10.1214/aos/1176347973 6.1

Friedman, J. H., and Bogdan, E. P. (2008). "Predictive learning via rule ensembles." *Annals of Applied Statistics* 2, no. 3, 916–954. DOI: 10.1214/07-AOAS148 4.2, 5

Friedman, J., Hastie, T. and Tibshirani, R. (2000). "Additive Logistic Regression: a Statistical View of Boosting." *Annals of Statistics* 28. DOI: 10.1214/aos/1016218223

Friedman, J. H., Hastie, T., and Tibshirani, R. (2008). "Regularized Paths for Generalized Linear Models via Coordinate Descent." Technical Report. Department of Statistics, Stanford University. 1.2, 4.2

Friedman, J.H. and Silverman, B.W. (1989). Flexible Parsimonious Smoothing and Additive Modeling, *Technometrics* **31**, no. 1, 3–21. DOI: 10.2307/1270359 6.1

"gbm." *CRAN - Package*. http://cran.r-project.org/web/packages/gbm 4, 4.5, 4.6

Hansen, L.K. and Salamon, P. (1990). "Neural network ensembles." *IEEE Transaction on Pattern Analysis and Machine Intelligence* **12**:10, 993-1001. 1

Hastie, T., R. Tibshirani (1985). Discussion of "Projection Pursuit" by P. Huber, *The Annals of Statistics* **13**, 502–508. DOI: 10.1214/aos/1176349528 6.1

Hastie, T., Tibshirani, R., and Friedman, J. (2001). *The Elements of Statistical Learning-Data Mining, Inference and Prediction*. Springer. 2.3, 2.4, 3.2, 3.3, 3.3.3, 3.3.4, 3.3.5, 3.11, 4.3.1, 5.1, 5.2.3

Hastie, T., Tibshirani, R., and Friedman, J. (2009). *The Elements of Statistical Learning: Data Mining, Inference, and Prediction, 2nd ed Edition*. Springer. DOI: 10.1007/BF02985802 2

Hinch, E. J. (1991). *Perturbation Methods*. Cambridge University Press. 4.1.2

Hinkley, D.V. (1969). Inference About the Intersection in Two-Phase Regression, *Biometrika* **56**, 495–504. DOI: 10.1093/biomet/56.3.495 6.1

Hinkley, D.V. (1970). Inference in Two-Phase Regression, *Journal of the American Statistical Association* **66**, 736–743. 6.1

Hjorth, U. (1989). On Model Selection in the Computer Age, *Journal of Statistical Planning and Inference* **23**, 101–115. DOI: 10.1016/0378-3758(89)90043-8 6.1

Ho,T.K.,Hull, J.J.,and Srihari,S.N. (1990)."Combination of Structural Classifiers." *Pre-Processings, International Association for Pattern Recognition Workshop on*

Syntactic & Structural Pattern Recognition, Murray Hill, NJ, USA, June 12–15, 123–136. 1

Ho, T.K. (1995). "Random Decision Forests." *3rd Intl. Conference on Document Analysis and Recognition*, 278–282. DOI: 10.1109/ICDAR.1995.598994 1.1

Huber, P. (1964). "Robust estimation of a location parameter." *Annals of Math. Stat* 53, 73–101. DOI: 10.1214/aoms/1177703732 3.3

"ipred." *CRAN - Package*. http://cran.r-project.org/web/packages/ipred 4.3

Kleinberg, E. (1990). "Stochastic Discrimination." *Annals of Mathematics and Artificial Intelligence* **1**, 207–239. 1

Kleinberg, E. (2000). "On the Algorithmic Implementation of Stochastic Discrimination." *IEEE Transactions on Pattern Analysis and Machine Intelligence* 22, no. 5, 473–490. DOI: 10.1109/34.857004

Ivakhenko, A. G. (1968). The group method of data handling—A rival of the method of stochastic approximation. *Soviet Automatic Control* **3**, 43–71. 1.1

lars. CRAN - Package. http://cran.r-project.org/web/packages/lars 3.3.5

Michie, D., Spiegelhalter, D. J., and Taylor, C. C. (1994). *Machine Learning, Neural and Statistical Classification*. New York: Ellis Horwood. 1, 1

Nisbet, R., Elder, J., and Miner, G. (2009). *Handbook of Statistical Analysis & Data Mining Applications*. Academic Press. 1

Owen, A. (1991). Discussion of "Multivariate Adaptive Regression Splines" by J. H. Friedman, *Annals of Statistics* **19**, 82–90. 6.1

Panda, B., Herbach, J., Basu, S., and Bayardo, R. (2009). "PLANET: Massively Parallel Learning of Tree Ensembles with MapReduce." *International Conference on Very Large DataBases*. Lyon, France. 4.3

Popescu, B.E. (2004). "Ensemble Learning for Prediction." Ph.D. Thesis, Statistics, Stanford University.

Quinlan, J.R. (1992). *C4.5: Programs for Machine Learning*. Morgan Kaufmann. 1, 2

"randomForest." *CRAN - Package*. http://cran.r-project.org/web/packages/randomForest/index.html 4.4

Rosset, S. (2003). "Topics in Regularization and Boosting." PhD. Thesis, Statistics, Stanford university. 3.3

rpart. CRAN - Package. http://cran.r-project.org/web/packages/rpart 2.1, 3.3.2

RuleFit. http://www-stat.stanford.edu/~jhf/R-RuleFit.html 5.2.1

Scholkopf, B., Burges, Burger, C.J.C. Burges, and Smola, A.J. (1999). *Advances in Kernel Methods Support Vector Learning*. MIT Press. 2, 4

Seni, G., Yang, E., Akar, S., Yield modeling with Rule Ensembles, *18th Annual IEEE/SEMI Advanced Semiconductor Manufactuting Conference*, Stresa, Italy, 2007. 5.3

Tibshirani, R. (1996). "Regression shrinkage and selection via the lasso." *J. Royal Statistics Society B.* 58, 267–288. 1.2, 3.3.3

Tibshirani, R., K. Knight (1999a). The Covariance Inflation Criterion for Adaptive Model Selection, *Journal of the Royal Statistical Society, Series B-Statistical Methodology*, **61** (pt. 3), 529–546. DOI:10.1111/1467-9868.00191 6.1

Tibshirani, R. and K. Knight (1999b). Model Search and Inference by bootstrap "bumping." *Journal of Computational and Graphical Statistics* **8**, 671–686. 6.1

Winsorize. http://www.itl.nist.gov/div898/software/dataplot/refman2/auxillar/winsor.htm 5.1

Wolpert, D. (1992). Stacked generalization. *Neural Networks* **5**, 241–259. DOI: 10.1016/S0893-6080(05)80023-1 1.1

Ye, J. (1998). On Measuring and Correcting the Effects of Data Mining and Model Selection, *Journal of the American Statistical Association* **93**, no. 441, 120–131. DOI: 10.2307/2669609 6, 6.1, 6.3, 6.2, 6.3

Zhao, P. and Yu, B. (2005). "Boosted lasso." *SIAM Intl. Conference on Data Mining.* Newport Beach, CA, 35–44.

Zou, H. and Hastie, T. (2005). "Regularization and Variable Selection via the Elastic Net." *SIAM workshop on Variable Selection.* Newport Beach, CA. DOI: 10.1111/j.1467-9868.2005.00503.x 1.2, 3.3.3

Authors' Biographies

GIOVANNI SENI

Giovanni Seni is a Senior Scientist with Elder Research, Inc. and directs ERI's Western office. As an active data mining practitioner in Silicon Valley, he has over 15 years R&D experience in statistical pattern recognition, data mining, and human-computer interaction applications. He has been a member of the technical staff at large technology companies, and a contributor at smaller organizations. He holds five US patents and has published over twenty conference and journal articles.

Giovanni is an adjunct faculty at the Computer Engineering Department of Santa Clara University, where he teaches an Introduction to Pattern Recognition and Data Mining class.

He received a B.S. in Computer Engineering from Universidad de Los Andes (Bogotá, Colombia) in 1989, and a Ph.D. in Computer Science from State University of New York at Buffalo (SUNY Buffalo) in 1995, where he studied on a Fulbright scholarship. He also holds a certificate in Data Mining and Applications from the Department of Statistics at Stanford University.

JOHN F. ELDER

Dr. John F. Elder IV heads a data mining consulting team with offices in Charlottesville, Virginia, Washington DC, and MountainView, California (www.datamininglab.com). Founded in 1995, Elder Research, Inc. focuses on federal, commercial, investment, and security applications of advanced analytics, including text mining, stock selection, image recognition, biometrics, process optimization, cross-selling, drug efficacy, credit scoring, risk management, and fraud detection. ERI has become the largest and most experienced data mining consultancy.

John obtained a BS and MEE in Electrical Engineering from Rice University, and a PhD in Systems Engineering from the University of Virginia, where he's an adjunct professor teaching Optimization or Data Mining. Prior to 15 years at ERI, he spent 5 years in aerospace defense consulting, 4 heading research at an investment management firm, and 2 in Rice's *Computational & Applied Mathematics* department.

Dr. Elder has authored innovative data mining tools, is a frequent keynote speaker, and was co-chair of the 2009 *Knowledge Discovery and Data Mining* conference, in Paris. His courses on analysis techniques – taught at dozens of universities, companies, and government labs – are noted for their clarity and effectiveness. John was honored to serve for 5 years on a panel appointed by the President to guide technology for National Security. His award-winning book for practitioners of analytics, with Bob Nisbet and Gary Miner – *The Handbook of Statistical Analysis & Data Mining Applications* – was published in May 2009.

John is a follower of Christ and the proud father of 5.

Printed in the United States
by Baker & Taylor Publisher Services